AF601240

Fundamentals of Disaster Management

The Authors

Dr. Sandip P. Nikam holds bachelor's degree in Agril. Engg. and masters degree in Soil and Water Conservation Engg. from Mahatma Phule Krishi Vidyapeeth, Rahuri. Maharana Pratap University of Agriculture and Technology, Udaipur awarded him Ph.D. degree with Gold Medal. He was worked with various internationally reputed NGOs engaged in Natural Resource Management and Climate Change. He has a special interest in the study of disasters and disaster management. Presently, he is working as an Assistant Professor at College of Agriculture, Dhule. He has 20 years of experience in NRM and teaching.

Prof. Atul B. Deshmukh holds a bachelor degree in Bachelor of Arts (English) and Masters degree in Library and Information Science from Savitribai Phule University of Pune, Pune. Earlier in 2008-09, he was working in National Chemical Laboratory and in July 2009 he is joined as a Librarian at College of Agriculture, Dhule. He has presented 7 research papers in National and International Conference. He has a special interest in the study of Library and Information Services (Online Services). Presently, he is working as a Librarian at College of Agriculture, Pune. He has 8 years of experience in Library Science field and teaching.

Dr. Ulhas Shantaram Surve is presently working as Associate Professor of Agronomy, Department of Agronomy, Post Graduate Institute, Mahatma Phule Krishi Vidyapeeth, Rahuri, Dist. Ahmednagar (MS). He is having 03 years B.Sc. (Agri.), 03 years M.Sc. (Agri.) in Irrigation Water Management and 06 years M.Sc. (Agri.) and Ph.D. in Agronomy teaching experience. He has guided two M. Sc. (Agri.) students in Irrigation Water Management as well as nine M. Sc. (Agri.) and two Ph. D. students in Agronomy. Act as SAC committee members for 38 students in M.Sc. (Agri.) and 5 Ph.D. students in various disciplines. He has published 21 research papers, 11 technical papers and 35 popular papers. He has participated in 05 short training courses, 02 winter schools and 11 seminars and workshops. He has awarded as The Best Citizens of India Award, 2012 from International Publishing House, New Delhi and Young Scientist Award, 2015 from IJTA and Serial Publication, New Delhi. He has published three practical manuals of M.Sc. and Ph.D. courses in Agronomy and two books. He has given four recommendations on micro-irrigation and fertigation. His research area of specialization is in micro-irrigation, fertigation, protected cultivation (colored shadents, polyhouse) and Integrated farming system.

Fundamentals of Disaster Management

Dr. Sandip P. Nikam

Prof. Atul B. Deshmukh

Dr. Ulhas S. Surve

2017

Scholars World

A Division of

Astral International Pvt. Ltd.

New Delhi – 110 002

© 2017 AUTHORS

Publisher's Note:

Every possible effort has been made to ensure that the information contained in this book is accurate at the time of going to press, and the publisher and author cannot accept responsibility for any errors or omissions, however caused. No responsibility for loss or damage occasioned to any person acting, or refraining from action, as a result of the material in this publication can be accepted by the editor, the publisher or the author. The Publisher is not associated with any product or vendor mentioned in the book. The contents of this work are intended to further general scientific research, understanding and discussion only. Readers should consult with a specialist where appropriate.

Every effort has been made to trace the owners of copyright material used in this book, if any. The author and the publisher will be grateful for any omission brought to their notice for acknowledgement in the future editions of the book.

All Rights reserved under International Copyright Conventions. No part of this publication may be reproduced, stored in a retrieval system, or transmitted in any form or by any means, electronic, mechanical, photocopying, recording or otherwise without the prior written consent of the publisher and the copyright owner.

Cataloging in Publication Data--DK
Courtesy: D.K. Agencies (P) Ltd. <docinfo@dkagencies.com>

Nikam, Sandip P., author.
Fundamentals of disaster management / Dr. Sandip P. Nikam, Prof. Atul B. Deshmukh, Dr. Ulhas S. Surve.
pages cm
Includes bibliographical references and index.
ISBN 978-93-86071-45-3 (International Edition)

1. Emergency management--India. 2. Emergency management--Government policy--India. I. Deshmukh, Atul B., author. II. Surve, Ulhas S., author. III. Title.

HV551.5.I4N55 2017 DDC 363.340954 23

Published by : **Scholars World**
A Division of
Astral International Pvt. Ltd.
– ISO 9015:2015 Certified Company –
4736/23, Ansari Road, Darya Ganj
New Delhi-110 002
Ph. 011-4354 9197, 2327 8134
E-mail: info@astralint.com
Website: www.astralint.com

Preface

India has been traditionally vulnerable to natural disasters on account of its unique geo-climatic conditions. Floods, droughts, cyclones, earthquakes and landslides have been recurrent phenomena. In 2014, 324 triggered natural disasters1 were registered. It was the third lowest number of reported disasters in the last decade, below the annual average disaster frequency observed from 2004 to 2013 (384). However, natural disasters still killed 7,823, a number largely below the annual average for years 2004-2013 (99,820), and 140.8 million people become victims worldwide, also below the 2004-2013 annual average (199.2 million). The loss in terms of private, community and public assets has been astronomical. Disaster management essentially deals with management of resources and information towards a disastrous event and is measured by how efficiently, effectively and seamlessly one coordinates these resources. The ability to effectively deal with disasters has become a challenge to modern technology. It is apparent that disaster problems cut across various disciplinary lines. One cannot effectively address disaster management a difficulty by focusing on the isolated problems of a single type.Effective disaster management is influenced by the activities of a host of independent organisations at national and inter-national level.

Disaster management at the individual and organisational level deals with issues of planning, coordination, communication and risk assessment. Accordingly, this book also covers these subjects to enhance their ability for better disaster response. For the most part, the case studies in this book deal with various disasters to promote discussion on issues of disaster and development, national policies on disaster management across the world and problems in disaster management. The conclusions of these case studies must be interpreted with caution; because, actual disaster situations always brings surprises and are unpredictable.

This book is intended to guide and to impart the fundamental knowledge for everyone interested in disaster and its management. It covers most of the topics related to disaster management. Special efforts have been put in to make the writing style simple. Instead of impressing the reader about literacy level of the author, prime importance has been given to reader friendliness and ease of understanding. This book has been divided into eight chapters covering the fundamental aspects like introduction to disaster, types of disaster, disaster management, disaster management and GIS-GPS-RS, physical and socio-economical impacts of the disaster. The facts and figures are provided in the Annexure. An effort has been made to cover the subject according to the syllabus of couple of Indian Universities who have taken lead in introducing the course on disaster management. We hope that the content of the book meets the requirement of these 'would be' disaster professionals. Disaster impact is often expressed in terms of the numbers of dead and injured and loss of property and resources. Therefore, the purpose of this book will be served if in the end it helps to minimize the economic losses and reduce injuries and death to the affected population.

The authors have freely drawn on the materials in various handbooks, publications, research reports, proceedings *etc.* This is the first attempt made to prepare this publication for the use of students, volunteers and peoples involved in the disaster management directly or indirectly.

A number of friends, teachers and seniors, too numerous to mention by name, have helped in bringing out this book. We are grateful to everyone of them. It is our pleasure to offer this book to the students.

Dr. Sandip P. Nikam

Prof. Atul B. Deshmukh

Dr. Ulhas S. Surve

Contents

Chapter 1

Introduction to Disaster

India is one of the most disaster prone countries of the world. It has had some of the world's most severe droughts, famines, cyclones, earthquakes, chemical disasters, rail accidents and road accidents. India is also one of the most terrorist prone countries. Natural disasters may occur at any time without warning and is considered dangerous for life and property. Even in the present age of technology development, perfect understanding of the causes of these disasters has not been achieved. Disasters were considered as inevitable, having little relationship with the social, economic or institutional context. As a result, the course of action followed was of emergency management and relief with welfare orientation. Interventions are mostly made after the disaster occurs, with the objective of returning the situation to, as it was before the event. People, who are forced to live in the same conditions, fall deeply in to the trap of poverty and become more vulnerable.

Responding to emergencies is no doubt an important aspect of disaster management. However, the absence of other import components such as disaster preparedness, risk management (risk identification, risk reduction and risk transfer) based on the root causes of the event and a sustainable approach towards relief and rehabilitation is a cause for concern.

There is a strong need to move from an emergency management culture to a culture of disaster preparedness for sustainable development of the country. The culture of inertia and fatalism must be overcome.

1.1 Disaster Risks in India

India is vulnerable, in varying degrees, to a large number of natural as well as man-made disasters. 58.6 per cent of the landmass is prone to earthquakes of moderate to very high intensity; over 40 million hectares (12 per cent of land) is prone to floods and river erosion; of the 7,516 km long coastline, close to 5,700 km is prone to cyclones and tsunamis; 68 per cent of the cultivable area is vulnerable to drought and hilly areas are at risk from landslides and avalanches. Vulnerability to disasters/emergencies of Chemical, Biological, Radiological and Nuclear (CBRN) origin also exists. Heightened vulnerabilities to disaster risks can be related to expanding population, urbanization and industrialization, development within high-risk zones, environmental degradation and climate change.

Disaster

The word disaster derives from the French word 'desastre'. The definition given by the Oxford English dictionary is **'anything that befalls of ruinous or distressing nature; a sudden or great misfortune, mishap or misadventure; a calamity'**. The generally accepted definition of disaster is an occurrence arising with little or no warning, which causes or threatens series disruption of life, and perhaps death or injury to large number of living beings and requires, therefore, a mobilization of efforts in excess of that normally provided by the statutory emergency services.

General Preparedness

The main characteristics of a major disaster are that irrespective of the origin, after a little while the scene is the same:

- ☆ Total chaos all around
- ☆ Lack of utilities – which we have always taken for granted
- ☆ No relief and rescue teams for several days
- ☆ Lack of medical facilities

Thus, the sufferings are not just due to the disaster, but, post-disaster, many more people die and suffer because of:

1. Lack of food, shelter
2. Lack of medical attention
3. Hygiene issues causing health hazards

1.2 Classification of Disasters

Major Classification

A) Classification Based on their Speed of Onset

Class	*Natural*	*Manmade*
Rapid on set	Earthquakes	Fires
	Flood	Technological accidents
	Hurricanes	Industrial accidents
	Volcanic Eruptions	Transportation accidents

Class	*Natural*	*Man Made*
	Land slides	
	Tsunamis	
Slow onset	Droughts	Famines
	Floods	Civil strife
	Epidemics	

B) Classification Based on Causes

Natural Disaster	*Manmade Disaster*
Floods	Nuclear disasters
Droughts	Chemical disasters
Cyclones	Biological disasters, Deforestation
Earthquakes	Building fire, Coal fire, Oil fire
Landslides	Air pollution, Water pollution
Avalanches	Industrial wastewater pollution
Volcanic eruptions	Road accidents, Rail accidents
Heat and Cold waves	Air accidents and Sea accidents

1.3 Categories of Disasters

A) Basic Category

1. Meteorological (Drought, Floods and Cyclones)
2. Topographical (Earthquakes and Volcanic eruption)
3. Accidental (Chemical accidents, Industrial Accidents and Oil spills)

B) Based on their Origin

The High Power Committee on Disaster Management (HPCDM) constituted in 1999, identified 31 various disasters and categorized them in to five major categories, primary based on their origin, these are:

1) Water and Climate Related Disasters

Floods, cyclones, tornadoes and hurricanes, hailstorm, cloud burst, heat wave and cold wave, snow avalanches, droughts, sea erosion, thunder and lighting.

2) Geographically Related Disasters

Landslides, earthquakes, dam failures/dam burst and mine fires.

3) Chemical, Industrial and Nuclear Related Disasters

Chemical and industrial disasters, nuclear disaster.

4) Accident Related Disasters

Forest fire, urban fires, mine flooding, oil spills, major building collapse, serial bomb blast, festival related disasters, electrical disasters and fires, air, road and rail accidents, boat cap zing and village fire.

5) Biologically Related Disaster

Biological disasters and epidemics and pest attack *etc.*

Chapter 2

Natural Disaster

These are primarily natural events. It is possible that certain human activities could maybe aid in some of these events, but, by and large, these are mostly natural events. A natural disaster is the effect of natural hazards that affects the environment and leads to financial, environmental and/or human losses. The resulting loss depends on the capacity of the population to support or resist the disaster, and their resilience. A disaster occurs when hazards meet vulnerability.

2.1 Floods

Floods are temporary inundation of large regions as a result of overflow of reservoir or of rivers flooding their banks because of heavy rains, high winds, cyclones, storm surge along coast, tsunami, melting snow or dam bursts.

Onset Type

Floods can happen gradually and take over hours, or can even happen suddenly due to heavy rains, breach of the water storage and control structures, spillovers, storm surge.

Warning

Except for flash flood there is usually a reasonable warning period. Heavy rain fall will give sufficient time to anticipate the occurrence of flood. In India, warning is issued by Central Water Commission; Irrigation and Flood Control Department, and Water Resource Department.

Vulnerability

Anything flood plains will get inundated. Buildings built of earth; weak foundations and water soluble materials will collapse and endanger humans and property. Basements of buildings are at risk.

Typical Effects

- Physical damage like landslides can happen, structures may damage due to high intensity flowing water.
- People and livestock deaths caused by drowning and outbreak of epidemics like diarrhea, viral infection *etc.*
- Water supplies can get contaminated.
- Food storage can be caused due to loss of entire harvest and spoiling of the stored grains due to flood water.
- Land can become infertile due to erosion, or can turn saline if sea water floods the area.

2.1.1 Flood Disaster Management/Mitigation Strategy

- Mapping of the flood plain is the primary step involved in reducing the risk of floods.
- Land use control will reduce danger of life and property when water inundates the flood plains and the coastal areas. Important facilities should be built in safe areas *i.e.* mainly on elevated land. In urban areas water holding structures likes ponds and lakes may be constructed.
- Construct structures in the flood plains to with stand flood forces and seepage. If necessary built structures on stilts and platforms.
- Flood control aims to reduce flood damage. This can be done by flood reduction which decreases the amount of runoff by treatment like reforestation, protection of vegetation, clearing of the debris from streams and other water holding areas, construction of ponds and lakes *etc.* Flood diversion includes reeves, embankments, dams and channel improvements. Dams can store water and can release water at a manageable rate. Flood proofing reduces the risk of damages. Measures include use of sand bags to keep flood water away, blocking or sealing of

doors and windows of the house *etc.* Houses may be elevated by raising them through structural means or by raising the land. Buildings should be constructed away from the water bodies.

2.1.2 During the Floods

1. Deployment of action group
2. Priority on rescue and evacuation operation.
3. Ensuring availability of dry food in isolated areas and cooked food in relief camps.
4. Ensuring adequate supply of potable water.
5. Ensuring availability of medical facilities like doctors, medicines, hospital beds *etc.*

2.2 Drought

Drought is a temporary reduction in water or moisture availability, significantly below the normal or expected amount for a specific period. This condition occurs either due to inadequacy of rainfall or lack of irrigation facilities, under exploitation or deficient availability for meeting the normal crops requirements in the context of agro climatic condition prevailing in any particular area. Rajasthan is the most drought prone state of India. For an area to be declared drought affected the Government uses two key indicators: first a significant reduction in the amount of rainfall received compared to the normal rainfall of the region; and second the anewari (estimate of crop production based on visual estimates and crop cutting experiments) falls to below 37 per cent of its normal value.

2.2.1 General Characteristics

Types of Droughts

Drought proceeds in sequential manner. Its impacts are spread across different domains as listed below:

1) Meteorological Drought

Meteorological drought is simple absence/deficit of rainfall from the normal. It is the least severe form of drought and is often identified by sunny days and hot weather.

2) Hydrological Drought

Hydrological drought often leads to reduction of natural stream flows or ground water levels, plus stored water supplies. The main impact is on water resource systems.

3) Agricultural Drought

This form of drought occurs when moisture level in soil is insufficient to maintain average crop yields. Initial consequences are in the reduced seasonal output of crops and other related production. An extreme agricultural drought can lead to a famine, which is a prolonged shortage of food in a restricted region causing widespread disease and death from starvation.

4) Socio-economic Drought

Socio-economic drought correlates the supply and demand of goods and services with the three above-mentioned types of drought. When the supply of some goods or services such as water and electricity are weather dependent then drought may cause shortages in supply of these economic goods.

2.2.2 Distribution Pattern

- ✰ Around 68 per cent of India's total area is prone to drought.
- ✰ 315 out of a total of 725 talukas in 99 districts are drought prone.
- ✰ 50 million people are annually affected by drought.
- ✰ In 2001 more than eight states suffered the impact of severe drought.
- ✰ In 2003 most parts of Rajasthan experienced the fourth consecutive year of drought.

2.2.3 Onset-Type and Warning

Drought is a slow onset disaster and it is difficult to demarcate the time of its onset and end. Nevertheless, a good time to take stock is to estimate the total rainfall at the end of the monsoon (usually October/November). Droughts adversely affect rain- fed crops to starts with and subsequently irrigated crops. Areas with minimum of alternative water resources to rainfall (ground and canal water supplies) areas subjected to drastic environmental degradation such as denuded forest lands and altered ecosystems and areas where livelihoods alternative to agriculture are least developed are most Vulnerable to drought. Hard men, landless laborers, subsistence farmers, women, children and farm animals are the most Vulnerable groups affected by drought conditions.

2.2.4 Typical Effects

- ✰ Drought, different from other natural disasters, does not cause any structural damage.
- ✰ Effects on loss of crop, dairy, timber and fishery production.
- ✰ Increase in energy demand for pumping water.
- ✰ Reduction in energy production.
- ✰ Increase in unemployment.
- ✰ Loss of biodiversity.
- ✰ Reduces water, air and landscape quality.

- ☆ Ground water depletion, food shortage, health reduction and loss of life.
- ☆ Increases poverty, reduced quality of life and social unrest leading to migration.

2.2.5 Drought Disaster Management

- ☆ As far as possible reduce, reuse and recycle water for domestic use, industrial use and agriculture.
- ☆ Implement water conservation and water harvesting measures like check dams, contour bunds, gabion structures, roof top rain water harvesting, farm ponds, percolation tanks *etc.*
- ☆ Repair leaking taps, tanks, pipes *etc.*
- ☆ Avoid using water intensive devices like flush toilets, sprinklers for lawns, bath tubs, washing of vehicles *etc.*

2.3 Cyclone

Cyclones are violent storms often of vast extent, characterized by high winds rotating about a calm centre of low atmospheric pressure. This centre moves onwards, often with a velocity of 50 km/hour.

2.3.1 Onset-Type and Warning

Cyclones strike suddenly although it takes time to build up. Satellite tracking can track the movement since the buildup and the likely path can be projected. Warning and evacuation is done along the projected path. It is difficult to predict the path of cyclone with accuracy. The Indian Meteorological Department issues warnings against severe weather phenomenon like tropical cyclones, heavy rains, and snow, cold and heat waves *etc.*

2.3.2 Typical Effects

- **Physical effect**: Structures will be damaged or destroyed by the wind force, flooding, storm surge and landslides, buildings and communication systems can be damaged. Roofs of light weight material suffer severe damage.
- **Causalities and public health**: Due to flooding the water supplies can get contaminated and lead to causalities, viral outbreaks, diarrhea and malaria.
- High winds and rains will ruin the standing crops and food stock lying in low lying area.

Cyclones in India generally strike the East Coast; some of the Arabian Sea cyclones strike the west coast of India as well, mainly the coasts Gujarat and North Maharashtra. Out of the storms that develop in the Bay of Bengal, over 58 per cent approach or cross the east coast in October and November.

2.3.3 Cyclone Disaster Management

Precautionary Measures

- Check houses, secure loose tiles.
- Remove dead or dying trees.
- Keep hurricane and torch ready for emergency.
- Demolish condemned building.

When Cyclone Threatens

- Keep transistor radio on.
- Don't be misled by rumors'.
- Get extra food, store extra drinking water.
- Get away from low lying areas.
- Move valuables, important papers and documents to upper floors.
- Remove cattle to safer places.
- Remain calm and meet emergency.

Post Cyclone Measures

- Remain in shelter until advised to return home.
- Get inoculated at the nearest hospital.
- Keep away from dangling wire from electric poles.
- Dispose of dead bodies and carcasses quickly as early as possible.

2.4 Earthquake

Earthquakes refer to shaking of earth. There is continuous activity going on below the surface of the earth. There are several large plates (size of continents) below the surface of the earth, which move (at a very slow speed). As a part of this movement, sometimes, they collide against each other. And, after the collision, they might still continue to push each other. As they continually keep pushing each other, there is a pressure building up – across these plates below the surface. And, then, at a certain time, one of the plates might slide over another. This causes an earthquake.

An earthquake is a sudden slipping or moment of a portion of the earth's crust or plates, caused by a sudden release of stresses.

The earth's outer shell is divided in to seven measure and some smaller plates, which are constantly in a dynamic state, pushing against, pulling away from, or grinding past one another Forces build up as the plates attempt to move in relation to each other. When the grips along the fault give away, stored energy are released in the form of earth tremors (earthquakes), volcanic activity *etc.*

2.4.1 Onset-Type and Warning

Earthquake is a sudden hazard. It occurs at any time of the year, day or night, with sudden impact and without any warning sign. Extensive research has been conducted in recent decades but there is no accepted method of earthquake prediction as on date.

2.4.2 Vulnerability

An earthquake may last for seconds or minutes, while aftershocks may occur for months after the main earthquake. India has had along history of earthquake occurrences. About 65 per cent of total area of the country is vulnerable to seismic damage of varying degrees.

Earthquake Zones

India is divided into the following five seismic zones:

1. Zone V: This is referred as very high damage risk zone.
2. Zone IV: This is referred as high damage risk zone.
3. Zone III: This is termed as moderate damage risk zone.
4. Zone II: This is termed as low damage risk zone.
5. Zone I: This is termed as very low damage risk zone.

The most vulnerable areas according to the present seismic Zone map of India are located in the Himalayan and sub Himalayan regions, North- East, Kutch and the Andaman and Nikobar Islands.

2.4.3 Typical Effects

- ☆ **Physical damage**- Damage and loss of buildings and service structures. Fire, floods may erupt due to dam failures, landslides could occur.
- ☆ **Causalities**- Often high, near the epicenter and in places where the population density is high and buildings are not resistant to earthquake forces.
- ☆ **Public health**- Multiple fracture injuries, moderately and severely injured is the most widespread problem. Breakdown in sanitary conditions and other unhygienic conditions could lead to epidemics.
- ☆ **Water supply**- Severely affected water supply due to failure of water supply distribution network and storage reservoirs. If supply lines of fire hydrants are vulnerable they could hamper fire service operations.
- ☆ **Transport network**- Severely affected due to damage to roads and bridges, railway tracks, airport runways and related infrastructure.
- ☆ **Electricity and communication**-All links affected, transmission towers, transponders, transformers may collapse.

2.4.5 Earthquake Disaster Management

- ☆ Move only if necessary to reach a safe place.
- ☆ If in a coastal area, move to higher ground. Earthquakes often generate tsunamis.
- ☆ Look for and extinguish small fires.
- ☆ Help neighbors who may require assistance.
- ☆ Keep Radio and TV on for emergency information and instructions.
- ☆ Don't be misled by rumors'.
- ☆ Inspect the home for damage. Aftershocks can cause additional damage to unstable buildings.
- ☆ Take photo graphs of the home and its contents for insurance purpose.

- ☆ Don't drink well or tube well water after an earthquake until it is tested scientifically. There may be poisonous gases or minerals in the water.
- ☆ Get inoculated at the nearest hospital.
- ☆ Dispose of dead bodies and carcasses as quickly as possible.

2.5 Avalanche (Land Movement Disaster)

2.5.1 Slipping of Ice or Snow on a Mountain

An avalanche is a rapid flow of snow down a slope, from either natural triggers or human activity. Typically occurring in mountainous terrain, an avalanche can mix air and water with the descending snow. Powerful avalanches have the capability to entrain ice, rocks, trees and other material on the slope. In mountainous terrain avalanches are among the most serious objective hazard to life and property, with their destructive capability resulting from their potential to carry on enormous mass of snow rapidly over large distances.

An avalanches are classified by their morphological characteristics such as the nature of the failure, the sliding surface, the propagation mechanism of the failure, the trigger of the avalanche, the slope angle, direction and elevation, Avalanche size, mass and destructive potential are rated on a logarithmic scale, typically made up to 4 to 7 categories, with the precise definition of the categories depending on the observation system of forecast region.

2.5.2 Occurrence of Avalanches (Onset Type)

1. When the stress on the snow exceeds shear, ductile and tensile strength either within the snow pack or at the contact of the base of the snow pack with the ground or rock surface.
2. Avalanches can occur in a standing pack Precipitated snow is to accumulate in to a snow pack during typically winter seasons and high altitudes.
3. Snow pack evaluation factors are heating by sun, radiation cooling and vertical temperature gradients in standing snow, snowfall amounts, and snow types.
4. Generally, mild winter weather will promote the settlement and stabilization of the snow pack, and conversely very cold, windy or hot weather will weaken the snow pack.
5. Any wind stronger than a light breeze can contribute to a rapid accumulation of snow on sheltered slopes downwind.
6. Snowstorms and rainstorms are important contributors to avalanche danger.
7. Day time exposure to sunlight will rapidly destabilize the upper of snow pack. During clear nights, the snow pack can strengthen or tighten, through the process of long- wave radioactive cooling.

2.5.3 Triggers

1. Natural triggers of avalanches include additional precipitation, radioactive and convicting heating, rock fall, ice fall and other sudden impact.
2. Human triggers of avalanches include skiers, snowmobiles, and controlled explosive works.
3. Avalanches can not only entrain additional snow but can also, give the sufficient accumulation of overburden due to smaller avalanches.

2.5.4 Avalanches Disaster Management

2.5.4.1 Prevention

There are several ways to prevent avalanches and lessen their power and destruction.

1. They are employed in areas, where avalanches pose a significant threat to people, such as sky resort and mountain towns, roads and railways.
2. Explosives are used to prevent avalanches, especially at sky resorts where other methods are often impractical.
3. Explosive charges are used to trigger small avalanches before to build large avalanches.
4. Snow fences and light walls can be used to direct the placement of snow.
5. The sufficient density of trees can greatly reduce the strength of avalanches.

6. Trees can either be planted or they can be conserved, such as in the building of a sky resort, to reduce the strength of avalanches.
7. Artificial barrier can be very effective in reducing avalanche damage. There are several types (a) Snow net-guy wires, these barriers are similar to those used for rock slides. (b) Snow fence (Rigid fence structure) - may be constructed of steel, wood or pre-stressed concrete and expensive.
8. The barriers of concrete, rocks or earth are placed right above the structure, road or railways; they are trying to protect or to channel avalanches.

2.5.4.2 Safety in Avalanche Terrain

- **Terrain management**- It involves reducing the exposure of an individual to the risk of travelling in avalanche terrain by carefully selecting what areas of slopes to travel on.
- **Group management**- It is practice of reducing the risk of having a member of a group or a whole group involved in an avalanche. Minimize the number of people on the slope, and mountain separation. Stop or camp only in safe locations. It is generally recommended not to travel alone, because there will be no one to witness your witness your burial and start the rescue.
- **Risk factor awareness**- It requires gathering and accounting for a wide range of information such as the meteorological history of areas, the current weather and snow conditions and physical and social indicators of the group.
- **Leadership**- It requires well defined decision making protocols that use the observed risk factors. The decision making frame works are taught in a variety of courses provided by resource centers in Europe and North America.

2.6 Landslide

What are Landslides?

Landslides are down slope transport of soil and rock resulting from naturally occurring vibrations, changes in direct water content, removal or lateral support, loading with weight and weathering, or human manipulation of water courses and slope composition.

Landslides vary in types of movement (falls, slider, topples, lateral spread, flows) and may be secondary effects of heavy storm, earthquakes and volcanic eruption. Landslides are more wide spread than any other geological event.

2.7 Volcanic Eruptions

A volcano is an opening, or rupture, in a planet's surface or crust which allows hot magma, ash and gases to escape from below the surface.

2.7.1 Occurrence (Onset) of Volcanoes

1. They are generally found where tectonic plates are diverging or converging. The Mid Atlantic Ridge, has examples of volcanoes caused by divergent tectonic plates pulling apart, the pacific ring of fire has examples of volcanoes caused by convergent tectonic plates coming together. By contrast, volcanoes are usually not created where two tectonic plates slide past one another.
2. Volcanoes can also form where there is strengthening and thinning of the earth's crust (called "non- hotspot intra plate volcanism") such as in the African Rift valley, and North America.
3. Volcanoes can be caused by mantle plumes. These so-called hotspots for example at Hawaii and found elsewhere in the solar system.

4. Conical mountain/hill emits molten rock/lava/gases through a fissure or vent in the crust of the earth.
5. More than 80 per cent of the earth surface comes from volcanoes.
6. Seismographic monitoring, tilt meters and surveillance by satellite all serve to predict activity in a volcano.
7. The world's largest active volcano is Mauna lava in Hawaii.

Lava can flow slowly or erupt violently in to the air. The rocks blown out of a volcano- called pyroclastic rocks - fall back to earth as dust, ash, cinder or pumice. Most volcanic ash falls to the ground; cemented together by water it forms a rock called volcanic tuff.

Depending on how often they erupt, volcanoes may be classified as 1) active 2) intermittent and 3) dormant or extinct.

- ☆ **Volcanic ashes-** Volcanic rock which is exploded from a vent in fragments less than an inch (2.5cm) in size. Volcanic ash particles are like small sharp glass particles that damage anything they come across.
- ☆ **Debris avalanches-** Debris that is transported away from the slope, due to the instability of the Volcanoes slope. The bigger avalanches, the greater its speed is more dangerous.
- ☆ **Landslides-** A gradual, down slope movement of mass of bedrock. The mixture of debris from a landslides or avalanches with water may produce harmful lahars.
- ☆ **Volcanic eruption-** it can precipitate natural disasters such as earthquakes, flash floods, acid rain and tsunamis.

2.7.2 Volcano Disaster Management (Safety Measures)

Precaution

Anyone living in the vicinity of a volcano should have a disaster supply kit prepared, including a pair of goggles and disposable breathing mask for each member of the family. Of course, it is advisable to stay away from active volcano sites.

If the volcano erupts:

- ☆ If possible, immediately leave the area.
- ☆ If caught near a stream, beware of mudflows. Do not cross bridge if a mudflow is approaching. They can move faster than you can walk or run.
- ☆ Avoid river valleys and low lying areas.
- ☆ Protect yourself from falling ash by wearing long sleeved shirts and long pants. Use goggles and wear a dust mask to help with breathing.
- ☆ Stay away from areas downwind from the volcano. It is best to stay indoors until the ash has settled. Close doors, windows and all ventilations in the houses (Chimney vents, furnaces, air conditioners, fan *etc.*)
- ☆ When the ash has settled, clear it from roofs and rain gutters.
- ☆ Avoid driving, volcanic ash can clog engines, damage moving parts, and stall vehicles, if you have to drive, keep speed down to 56 km/hr or slower.

Believe it or not, there are advantages to living near a volcano. Volcano provides geothermal resources which are converted in to energy. Moreover, when a volcano erupts it throws out a lot of ash. Although initially this ash is very harmful to the environment, in the long run, the ash layer will turn in to extremely fertile soil and rich in minerals. And tourists flock to volcano sites because the sunsets and views that they can breathtakingly beautiful.

2.8 Cold and Heat Waves

2.8.1 Cold Waves

A cold wave is a weather phenomenon that is distinguished by a cooling of the air. A cold wave is a rapid fall in temperature within a 24 hours period requiring substantially increased protection to agriculture, industry, commerce and social activities. The precise criterion for a cold wave is determined by the rate at which the temperature falls and the minimum to which it falls. A cold wave can cause death and injury to livestock and wildlife. Exposure to cold mandates greater caloric intake for all animals, including humans and if a cold wave is accompanied by heavy and persistent snow, grazing animals may be unable to reach need food and die of hypothermia on starvation.

2.8.1.1 Typical Effects

1. Extreme winter cold often causes poorly insulated water, pipelines and mains to freeze.

2. Poorly protected indoor plumbing ruptures as water expands within them, causing much damage to property and costly insurance claims.
3. Demand for electrical power and fuels rises dramatically during cold waves.
4. Failure of the transportation *i.e.* motor vehicles.
5. Fires become even more of a hazard during extreme cold. The air during a cold wave is denser and draws a more intense fire, because denser air contains more oxygen.
6. Cold waves that bring unexpected freezes and frosts during the growing season, which can kill plants. Such cold waves have caused famines.

2.8.1.2 Counter Measures

1. Extreme cold requires that fuel powered machinery to be used even part time must be run continuously. Example Siberia.
2. Internal plumbing can be wrapped and water continuously runs through pipes.
3. The homeless may be arrested and taken to shelters, schools and other buildings can be converted in to shelters.
4. Hospitals can prepare for the admission of victims of frost bite and hypothermia.
5. Suitable stocks of food and forage can be secured before cold waves for people and animals (live stocks) respectively.
6. Smudge pots can bring smoke that prevents hard freezes on a farm.
7. Vulnerable crops may spray with water that protect the plants by freezing and absorb the cold from surrounding air.
8. People can also stock candles, match boxes, flashlights, fuel and stoves.

2.8.2 Heat Waves

Definition

It can be defined as a prolonged period of excessive heat. In Australia, a heat wave is defined by appropriate heat index with an acceptable event threshold and duration and related it to the climatologically area. In the United States, heat waves are the second greatest cause of human mortality resulting from a natural hazard, killing more people.

2.8.2.1 Heat Wave Threat (Impact of Heat Waves)

1. High temperatures increase hospital admission relating to heat stress, dehydration.
2. High temperatures increase rate of certain crimes particularly those related to aggressive such as homicide.

3. High temperatures increase number of work related accidents and reduced work productivity and decreased sports performance.
4. High temperatures can also cause significant economic losses through livestock/crop losses and damage to roads, railways, bridges, power reticulation, infrastructure and electrical equipment.
5. Heat wave conditions also lead directly to significant increase in demand for electricity to power domestic air conditioners, water consumption and retail sales of cold drinks.

2.8.2.2 Heat Wave Risks

The level of heat discomfort is determined by a combination of factors-

- ☆ Meteorological- air temperature, humidity, wind and direct sunshine.
- ☆ Cultural- Clothing, occupation and accommodation, and
- ☆ Physiological- Health, fitness, age and level of acclimatization.

In heat stress, body needs to keep our inner body temperature close to 37°C. The body responds to this stress through three stages.

1. Heat cramps- Muscular pains and spasms caused by heavy exertion. It is least serve stage. There is signal; the body is having trouble with heat.
2. Heat exhaustion-typically occurs, when people exercise heavy work in a hot, humid place, where body fluids are lost through heavy sweating. Blood flow increases to the skin and decrease to the vital organs. This results in mild shock with symptoms of cold, clammy and pale skin, fainting and vomiting. If not treated, the victim may suffer from heat stroke.
3. Heat stroke- is life treating. The victim's temperature control system, which produces sweating to cool the body, stops working. The body temperature may exceed 40.6°C causing brain damage and death if the body is not cooled quickly.

2.8.2.3 Counter Measures

There have been significant improvements to decrease the vulnerability of the population through:

1. Use of air conditioners
2. Better housing design
3. Better clothing
4. A trend towards more people working indoors
5. Education in indoors
6. Prediction of temperature forecasts to seven days

2.9 Global Warming

2.9.1 What is Global Warming?

Global warming is defined as the increase of the average temperature on earth. As the earth is getting hotter, disaster like hurricanes, droughts and floods are getting more frequent.

Over the last 100 years, the average temperature of the air near the earth's surface has risen little less than 1°C (0.74±0.18°C, or 1.3±0.32°farenheit). Does not seem all that much? It is responsible for the conspicuous increase in storms, floods and raging forest fires we have seen in the last ten years, though, say scientist.

Global warming is the increase in the average temperature of Earth's near surface air and oceans, since the mid- 20th century and its projected continuation. Global surface temperature increased 0.74±0.18°C (1.33±0.32°F) between the start and the end of the 20th century. Intergovernmental Panel on Climate Change (IPCC) concludes temperature increase since the middle of the 20th century was likely caused by increasing concentrations of greenhouse gases resulting from human activity such as fossil fuel burning and deforestation. The IPCC also concludes that variation in natural phenomena such as solar radiation and volcanic eruptions had a small cooling effect after 1950.

Climatic model projections summarized in the latest IPCC report indicate that the global surface temperature is likely to rise a further 1.1 to 6.4°C (2.0 to 11.5°F) during the 21st Century. However, warming is expected to continue beyond 2100 even if green house gas emissions stop, because of the large heat capacity of the oceans and the long life time of carbon dioxide in the atmosphere.

2.9.2 Climate Change and the Environment

We've seen how the water cycle is affected by global warming, which has forced us to change the way we live. But, what about the natural environment? In this topic, we'll explore the environmental impacts of diminishing streams and groundwater levels as well as increases in ocean temperature. The serious consequences of ice melts will also be discussed and how this can significantly accelerate climate change even further. More evaporation than rainfall means many fresh water bodies in hotter climates have been dropping to record lows. This includes rivers and streams around the world. Many streams that flowed all year round now only flow a few months a year or weeks. And this has had devastating effects on dependent ecosystems.

Rivers and streams are vital, because of their riparian zones. These zones are located along the banks, where the water meets the land. It's here that some of the most biodiverse ecosystems exist. Not only do streams provide water for animals to drink, it's on the banks of flowing streams that insects can breed. Insect nourish birds, fish, and smaller reptiles, who in turn sustain smaller mammals. What's happening now is perennial streams are becoming seasonal or drying up completely. So the more complex, larger insects that have longer life cycles don't have time to breed and hatch. With less insects, fish and smaller animal populations also drop. The banks also hold riparian vegetation, which is especially edible and nutritious.

This supports large plant-eating mammals. And with large plant-eating mammals, carnivorous animals can hunt and survive. Less stream flow means less biodiversity. The Okavango Delta in Botswana supports the largest number of carnivores in the world and is visited by the highest number of migratory land mammals, all reliant on the annual reins of the Angolan highlands, several hundred miles north of the delta. These rains flood the patch desert rivers of the Kalahari Desert, providing the nourishing riparian vegetation that the animals rely on to get through the dry season. When animals arrive here, they have survived a severe summer and are living on a knife's edge. Sensitive ecosystems like this are at risk as the impact of climate change means they have to wait longer and longer for the floods to arrive. Without rainfall, groundwater levels drop and soils dry up. With global warming, this has been more intense in areas with drier climates. The northern Jarrah Forest of Western Australia is a significant example of what happens when groundwater drops. There's no the forest like it on earth. It was only a few years ago that this evergreen forest started dying in large patches. It was the most extensive tree collapse this region has ever recorded. The cause of the tree deaths was identified as drought. Since the '70s, rainfall in this area has been declining in distinct stages, to the point where now water bodies only receive 30 per cent of the water they used to. As groundwater levels drop, tree root systems can no longer access water. And so in the summer of 2010, they died on a mass scale.

Groundwater also supports stygofauna, which are small living creatures that reside in groundwater itself. They tend to be blind and transparent, mainly crustaceans, but can also be snails, beetles, mites, worms, and fish. Stygofauna are important in filtering groundwater and improving its quality before water reaches lakes and oceans. They feed on rotting matter that contain nutrients and maintain the flow of groundwater to surface water by keeping the soil porous and clear from accumulated contaminants. While these animals improve water quality, they are still very sensitive to changes in quality like salinity and nutrients. They are also highly sensitive to water level fluctuations. A groundwater drop of up to 10 metres has been seen in stygofauna habitats since the beginning of the century. With the increase of salinity from sea level rises, stygofauna are especially vulnerable. Oceans regulate the climate. They're giant temperature stabilisers for the earth. The increasing heat in the atmosphere has brought about rises in ocean temperature and sea levels from melting ice caps. Oceans are approximately half a degree hotter than they were at the beginning of the 20th century. This may not sound like much, but for sea life, this is a monumental change. By 2050, if patents continue, it's predicted oceans will be another one degree hotter and 30 centimetres higher, impacting sea life in a number of different ways.

Marine animals that require cooler temperatures migrate to deeper or colder waters that may have been unlivable before. For animals that cannot move, however, the situation becomes precarious. Coral reefs are an example of this. Coral anemones are absolutely vital and sustain the most biodiverse and populous marine life in the oceans. Corals are in fact living animals that form reefs over many decades, even thousands of years. They form reefs in places that have ideal temperatures, a result of enough sunlight at specific depths. When these conditions change, with either rising sea levels or prolonged high temperature, corals release a stress response by releasing this symbiotic algae, which are the components responsible for photosynthesis and building the coral's bony structure. The result is coral that looks white and skeleton-like. This coral bleaching, as it's called, is now a common trait with coral reefs around the world, including the largest in the world, the Great Barrier Reef. It takes years for them to recover. But in the meantime, the vibrant life they attract deplete. And the bleached coral themselves are prone to death and disease. With the dismal predictions of global warming, coral bleaching is expected to be a more frequent and severe phenomenon. Experts have been closely analysing ways climate change itself can reach a threshold and accelerate itself in what is termed as runaway climate change.

The melting of polar ice sheets doesn't just cause a rise in sea levels. But it also limits albedo. Albedo is the amount of the sun's light and heat the earth reflects back into space. The darker the earth, the more light it absorbs and the hotter it will get. When the white-reflective ice sheets melt, they expose the darker land and sea which will absorb heat from the sun, further heating up the earth and accelerating global warming. So not only does the earth lose more of its cooling reflective capability, but it becomes more of a heat sponge. In the summer period of 2012, satellite imagery captured 97 per cent of the Greenland ice sheet thawing, which has never been seen before. Polar ice sheets had been retreating year after year the last two decades, and it has seen albedo drop from 75 per cent in 2000 to 68 per cent in 2012. And that's not the only way climate change can accelerate itself. Locked away in Northern Siberia's forests is 760 million hectares of carbon-containing permafrost. Permafrost covers almost a quarter of the northern hemisphere. It is packed with stored greenhouse carbon, waiting to be released upon the ice melting. In Northeastern Siberia, ice lakes full of methane gas exist. And these lakes have shown to catch fire when ignited. The methane from the permafrost here leaks naturally and equates to 68 per cent of atmospheric carbon. You can imagine how serious it is if global warming fully impacts this region. Not only is the packeted with greenhouse gases, but it's packed with methane, a greenhouse gas that has 25 times more heat generating power in CO_2. A series of methane fires have already occurred this century, instigated by the melting permafrost. These fires burnt an enormous 20 million hectares of forest, an area the size of Great Britain, and released unprecedented amounts of stored carbon into the atmosphere.

2.9.2 Sea Level Rise

An increase in global temperature will cause sea level to rise and will change the amount and pattern of precipitation, probably including expansion of subtropical deserts. Warming is expected to be strongest in the Arctic and would be associated with continuing retreat of glaciers, permafrost and sea ice, other likely effects include changes in the frequency and intensity of extreme whether events, species extinctions and changes in agricultural yields, warming and related charges vary from region to region around the globe, though the nature of these regional variation is uncertain.

2.10 Ozone Depletion

Definition- Destruction of ozone in the ozone layer attributed to the presence of chlorine from manmade Chloro-Fluro-Carbon (CFC) and other forces. The layer is thinning because ozone is being destroyed at a faster rate than it is being regenerated by natural forces.

- ☆ Depletion of the ozone layer means that for a given vertical path from the earth's surface to the sun, less ozone is encountered along the way with each passing year.
- ☆ This allows more UV-B to pass through the earth's surface, more damage to the DNA of all surface life and soon. Ozone is unique in its ability to absorb UV-B that is emitted by sun.
- ☆ Depletion is a long term decrease in ozone concentrations. It has nothing to do with the daily pluses and minuses of ozone concentrations. It is forcing function that destroyed ozone before it can absorb UV-B from the sun, and there are a few primary conditions for ozone depletes.
- ☆ Depletion is not what makes the "ozone hole" it is what makes the ozone hole larger; Ozone requires the following components for formation. Oxygen, UV light of 215 nm wavelength or shorter. Some nitrogen compounds also assist in forming ozone, even without VOCs being present.
- ☆ Ozone is decreased by time, temperature, water vapor, molecules that are oxidized by ozone and molecules that catalyze ozone destruction. Water vapor also serves to block the nitrogen assistance path of ozone formation.
- ☆ Any increase in destroyers of ozone or reduction in progenitors of ozone will serve to decrease the amount of ozone present at any particular place.

Chapter 3

Man Made Disasters

3.1 Nuclear Disasters

Definition

A nuclear and radiation accidents are usually defined as a loss of control of radioactive material, potential to cause radiation poisoning.

Nuclear Accidents

Are much larger in magnitude of effects than a typical radiation accident (defined by the International Atomic Energy Agency's International Event scale). The prime event of a "major nuclear accident" is one in which a reactor one is damaged and large amounts of radiation released, such as in the **Chernobyl Disaster** in 1986 (Ukraine and Russia).

Chernobyl Disaster

This occurred in 1986 in Ukraine and Russia. That accident killed 50 people directly and may cause as many as 4000 additional cases of fatal cancer over time, as well damaging almost $ billion of property.

A Criticality Accidents (Excursion or Power Excursion)

This occurs when a nuclear chain reaction is accidently allowed to occur in fissile material, such as enriched uranium or plutonium. The Chernobyl accident is an example of a criticality accident.

Transport Accidents

It can cause a release of radioactivity resulting in contamination or shielding to be damaged resulting in direct irradiation.

Human Error

It has been responsible for some accidents, such as when a person miscalculated the activity of a teletherapy source.

Enriched Uranium or Plutonium Nuclear Material

These are capable of supporting a self sustaining neutron chain reaction, here in after called nuclear criticality. The products of nuclear criticality are heat, radiation and radioactive materials called fission products.

3.2 Chemical Disaster

Chemical accident a cause large scale damage to an environment and or injures to or death of many animals, plants or humans.

Chemical disaster releases a quantity of toxic chemical in to the environment, resulting in death or injury to workers or members of nearby communities. Examples include the mercury waste poisoning of fish, resulting in 111 human deaths at Minamata, Japan; the release of methyl isocyanides from a chemical plant in Bhopal, India at a cost of 2000 lives; and a nuclear accident at Chernobyl, Ukraine requiring removal of 1,60,000 people from their homes.

The Bhopal disaster was an industrial catastrophe that took place at a pesticide plant owned and operated by Union Carbide (UCIL) in Bhopal, Madhya Pradesh, India. Around midnight on the intervening night of December 2-3, 1984, the plant released Methyl Isocyanides (MIC) gas and other toxins, resulting in the exposure

of over 500000 people. Estimates vary on the death toll. The official immediate death toll was 2,259 and Government of Madhya Pradesh has confirmed a total of 3,787 deaths related to the gas release.

Some 25 years after the gas leak, 390 tons of toxic chemicals abandoned at the UCIL plant continue to leak and pollute the groundwater in the region and affect thousands of Bhopal residents who depends on it. There are civil and criminal cases related to the disaster on going in the United States District Court, Manhattan and the District court of Bhopal, India against Union Carbide.

The UCIL factory was established in 1969 near Bhopal. About 50.9 per cent was owned by Union Carbide Corporation (UCC) and 49.1 per cent by various Indian investors including public sector financial institutions. It produced the pesticide (trademark Sevin) in 1979 a Methyl Isocyanides (MIC) production plant was added to the site. MIC an intermediate in carbaryl manufacture was used instead of less hazardous but more expensive materials. UCC understood the properties of MIC and its handing requirements.

During the night of December 2-3, 1984 large amounts of water entered tank containing 42 tones of Methyl Isocyanides. The resulting exothermic reaction increased the temperature inside the tank to over 200°C (392°F), raising the pressure to a level the tank was not designed to withstand. This forced the emergency venting of pressure from the MIC holding tank, releasing a large volume of toxic gases in to atmosphere. The reaction spread up because of the presence of Iron in corroding non stainless steel pipelines. A mixture of poisonous gases flooded the city of Bhopal, causing great panic as people woke up with a burning sensation in their lungs. Thousands died immediately from the effect of the gas and many were trampled in the panic.

Factors leading to the gas leaks include:

- ☆ The use of hazardous chemicals (MIC) instead of less dangerous ones.
- ☆ Storing these chemicals in large tanks instead of over 200 steel drums.
- ☆ Possible corroding material in pipelines.
- ☆ Poor maintenance after the plant ceased production in the early 1980s.
- ☆ Failure of several safety systems (due to poor maintenance)
- ☆ Safety system being switched off to save money- including the MIC tank refrigeration system which alone prevents the disaster.

Analysis shows that the parties responsible for the magnitude of the disaster are two owners, Union Carbide Corporation and the Government of India, and to some extent, the Government of Madhya Pradesh.

Typical Effects

1. Apart from MIC, the gas could contain phosgene, hydrogen cyanide, carbon monoxide, hydrogen chloride and carbon dioxide *etc.* produced in the storage tank or in the atmosphere.
2. The gas is denser and stayed close to the ground and spread outwards. The initial effects of exposure were coughing, vomiting, severe eye irritation and a filling of suffocation.
3. It causes large scale damage to environment, injure people, animal and plants.
4. It also causes damage or permanent injuries to death of animals, plants and humans.
5. The stillbirth rate increased by up to 300 per cent and neonatal mortality rate by 200.

3.3 Biological Disaster (Bio-terrorism)

Utilization of organisms' causing smallpox and anthrax by such terrorist groups can cause greater harm and panic. Biological agents are living organisms' or their toxic products that can kill or incapacitate people, livestock's and plants.

Bioterrorism can be defined as the use of biological agents to cause death, disability or damage mainly to human beings. Thus bioterrorism is a method of terrorist activity to prevail mass panic and slow mass causalities. The three basic groups of biological agents, which could be used as weapons, are bacteria, viruses and toxins. Biological agents can be dispersed by spraying them in to the air, by infecting animals that carry the disease to humans and by contaminating food and water.

Causes and Method of Delivery

Causes

- ☆ Biological weapons are potentially more powerful agents to mass casualties leading to civil disruptions.

- ☆ Any biological material which fulfills some of the criteria of bio weapons can utilize for widespread attention and to harm a selected target.
- ☆ Biological agents can be disseminated with readily available technology.
- ☆ Common agricultural spray devices can be adopted to disseminate biological pathogens of the proper particle size to cause infection in human population over great distances.
- ☆ The perpetrators can use natural weather conditions such as wind and temperature inversions *e.g.* ventilation system or air movement related to transportation *e.g.* subway cars passing through tunnels to disseminate these agents and thus infect or intoxicate a large number of people.
- ☆ The expense of producing biological weapons is far less than that of other weapon systems.

Methods of Dissemination and Delivery

Aerosols- biological agents are dispersed in to the air, forming a fine mist that may drift for miles. It may cause epidemic diseases in human beings or animals.

- ☆ Animals- some diseases are spread by insects and animals, such as fleas, mice, flies, mosquitoes and livestock.
- ☆ Food and water contamination- some pathogenic organisms' and toxins may persist in food and water supplies.
- ☆ Person to person- spread of a few infections agents is also possible.
- ☆ Humans- have been the source of infection for smallpox, plague and the lassa viruses.

The biological agents have potential enough to cause mass causalities. The very brief description of these biological agents is given as below.

Anthrax

The disease anthrax is caused by the germ-positive, non motile bacillus anthrax is anthrax makes an ideal biological weapon. Different clinical forms of the disease are observed, depending on the routs of exposure.

Smallpox

If used as a biological weapon, smallpox represents as a serious threat to civilian population. The fatality rate of 30 percent or more among unvaccinated persons. Smallpox virus is a member of genus orthodox virus, and it is closely related to the viruses causing cowpox, vaccinated and monkey pox. It is one of the largest DNA viruses known. Transmission of this virus can occur in several different ways generally by droplets occasionally by aerosol by direct contact with secretions or lesions from a patient. Smallpox scabs remain infectious until they fall off, whereas chick pox is no longer infections once the lesions are crusted.

Prevention and Mitigation Measures: General Measures of Protection

1. The general population should be educated and made aware of the threats risks associated with it. Only cooked food and boiled/chlorinated/filtered water should be consumed.
2. Clinical isolation of suspected and confirmed cases is essential.
3. Specialized laboratories should be established for a confirmatory laboratory diagnosis.
4. Existing disease surveillance system as well as vector control measures have been pursed more rigorously.
5. Mass immunization programme in the suspected area has to be followed up.
6. Enhancing of knowledge and skills of clinicians plays a vital role in controlling the adverse impact the attack.

3.4 Fire Disasters

3.4.1 Building Fire

Fire is a part of our life. When handled in correct manner, it is a workable tool in the kitchen, garden, factory, laboratory and jungle. If not handled properly, the same beautiful fire turns in to an inferno, which may kill damage or destroy people and property in to time. Therefore, it is necessary to learn about fire and take care before it goes out of control.

Concepts and Parameters of Severity

- ☆ **Fire is fast -** In less than 30 seconds a small flame can get completely out of control and turn in to a major fire. It only takes minutes for thick black smoke to fill a house. In minutes, a house can be engulfed in flames. Most of fires occur in the home when people are asleep. If you wake up to a fire, you won't have time to grab valuables because fire spreads too quickly and the smoke is too thick. There is only time to escape.
- ☆ **Fire is hot -** Heat is more threatening than flames. A fire that alone can kill. Room temperatures in fire can be 100 degree at floor level and rise to 600 degree at eye level. In healing this super hot air will scratch your lungs. This heat can melt clothes to your skin. In five minutes a room can get so hot that everything in it ignites at once. This is called flashover.
- ☆ **Fire is dark** - Fire is not bright; it is pitch black. Fire starts bright, but quickly produces black smoke and complete darkness. If you wake up to afire you may be blinded, disoriented and unable to find your way around the home, you've lived in for years.
- ☆ **Fire is deadly** – Smoke and toxic gases kill more people than flames do. Fire uses up the oxygen you need and produces smoke and poisonous gases that kills. Breathing even small amounts of smoke and toxic gases can make you drowsy, disoriented and short of breath. The odourless,

colourless fumes can fill you in to a deep sleep before the flames reach your door. You may not wake up in time to escape.

Vulnerability-Mitigation Measures

- ☆ To prevent a fire from occurring is more difficult than extinguishing a fire.
- ☆ In order to minimize the damage caused by fire, the fire fighting Department of Local Self Governments are actively promoting the review of fire prevention equipment in new and existing residential and commercial buildings.
- ☆ In residential fire, the number of elderly people who failed to escape and lose their lives is increasing yearly.
- ☆ It is our urgent task to precisely analyze the state of damage caused by fire, to install fire prevention equipment, and to adopt measures fire control, which minimize damage.
- ☆ The use of fire is indispensable for our daily life.
- ☆ The careless in using fire, may lead to a disaster, which damages or harms people and property.

Disaster Management (Measures)

- ☆ In the event of a fire, escape plans help you get out of your home quickly. Immediately leave the place.
- ☆ While escape through smoke, remember to crawl low, and keep your mouth covered. The smoke contains toxic gases which can disorient you.
- ☆ Designate one person to go to a neighbor's home to phone the fire department.
- ☆ Escape first and then call for help.
- ☆ Take the safest exit route to escape from every room.

- Every home should have at least one working smoke alarm to protect you and your family.
- Overheating, unusual smells, short circuits and sparks are all warning signs that appliance need to be shut off then replaced or repaired.
- Unplug appliances when not in use.
- Use safely caps to cover all unused outlets, especially if there are small children in the home.
- Don't come back to house/office till the fire brigade allows you.
- Once out, stay out.

3.4.2 Coal Fire

The coal fire refers to a burning or smoldering coal seam, coal storage pile or coal waste pile. The absorption of oxygen by a coal, which resulting in oxidation, is an exothermic reaction. This leads to an increase in temperature. If the temperature exceeds approximately 80°C, the coal can ignite and start to burn. This process referred as "spontaneous combustion", is the most common cause for coal fires of large extent. Mining operations expose formally covered coal to oxidation processes and additionally lead to the accumulation of large coal waste and storage piles. Coal fires can be ignited by lightning, forest or peat fires, mining accidents or careless human interaction.

Classification of Coal Fires

Depending on cause, location, age and burning intensity, coal fires can be classified in to different types. Coal fires can be human induced, when coal seams in areas of mining activity start to smolder or burn due to anthropogenic causes. They can occur naturally, when the fire is a result of natural spontaneous combustion. Furthermore underground and surface coal fires, cold and hot coal fires, paleo and arecent coal fires, as well as coal fires in coal seams, coal waste or coal storage piles can be differentiated.

Others additionally classify shallow (0 to 10 meters), intermediate (10-30m) and deep fires (> 30m), depending on their location underground. The oldest coal fire on a global scale burning for 2000 years located in New South Wales, Australia.

Coal Fires in China

Today, the main coal fire areas stretch along the coal mining belt in China, which extends for 5000 Kilometers (Km) from east to west along the north of country. Here more than 50 coal fields affected by coal fires have been identified. At present in China an estimated 20-30 million tons of coal burn each year. The economic loss of the valuable resource in China is the biggest problem, both in terms of the special area affected and the amount coal lost each year.

Beside the economic loss, coal fires pose many environment threats. The fires produce large amounts of green house-relevant and partially toxic gases including carbon dioxide (CO_2), methane (CH_4), and nitrogen oxides (NO_x), nitrous oxide

(N_2O), carbon monoxide (CO), and sulfur dioxide (SO_2). The coal fires lead to the degradation of water sources and agricultural areas.

Measures

1. Analyze the map and monitor the coal fire regions.
2. Satellite remote sensing can play a key role for coal fire detection and monitoring.
3. Remote sensing is a powerful tool for the generation of land cover and coal fires maps.
4. Relative and thermal indicators can be used to develop early warning or risk assessment strategies.

3.4.3 Forest Fire

A forest fire is a natural disaster consisting of a fire which destroys a forested area and can be a great danger to people to people, who live in forests as well as wild life. Forest fires are generally started by ignitening but also human negligence or arson, and can burn thousands of square Kilometers.

Forest fire also known as wild fires, vegetation fire, grass fire, or bush fire is common in areas of Australia, South Africa, U.S. and Canada, where climates are sufficiently to allow the growth of trees but feature extended hot and dry periods.

A Forest fires are caused by the drying out of branches leaves and therefore becomes highly flammable.

Fires can do weird things including:

- Crawing - spreads from bush to bush.
- Crown spread at an increditable pace through the top of the forest.
- Jumping/spotting burning branches and leaves carried away by wind.

To prevent fires there are large firefighter services including planes, fire trucks as well as small extinguishers depending on the severity of the fire.

3.5 Pollution

3.5.1 Air Pollution

Air pollution is the introduction of chemicals, particular matter or biological materials that cause harm or discomfort to humans or other living organism or damages the natural environment in to the atmosphere.

The Top-20 List

There is indeed a credible top-20 list for most-polluted cities. It was assembled by the World Health Organization in 2014 (ANNEXURE – V). It compares cities based on how much fine particulate matter they have in the air. This measurement is called PM 2.5, standing for particulate matter that is 2.5 microns in width or smaller. (There are about 25,000 microns per inch, so these particles are very small.)

Such pollutants at the ground level pose such immediate health risks as eye, nose, throat and lung irritation, and they can elevate the risk for lung and heart ailments over the longer term. These small particles typically come from exhaust from cars, trucks, buses and other vehicles, though they can also come from natural sources such as forest and grass fires.

Stratospheric Ozone Depletion

Due to air pollution has long been recognized as a threat to human health as well as to the earth's ecosystems. From cigarette smoke to global warming, air pollution has many different causes and affects us in many different ways. An awful lot of pollution is highly toxic and harmful to health, for example – That garden bonfire smoke contains over 350 times as much of the cancer causing chemical benzpyrene as a cigarette smoke. The car exhausts or garden bonfires, rotting food on landfills, forest fires caused by accident or arson, or fumes from factories are different types of air pollution.

Acid Rain

It is the term given to any precipitation that is above normal acidity, this includes snow and fog. The acidity or causticity of a solution is measured in pH. A neutral pH is 7, acidic is lower than 7 and a base is greater than 7. The normal pH of rain water is 5.5 (sightly acidic). The very highly acidic rain (pH=2.0) kills plants and animals and destroys buildings. Then air pollution hasn't gotten any better.

Acid rain occurs when the moisture in the air mixes with carbon, nitrogen and sulphur gases. These gases are released in to air by the burning of fossil fuels (natural gas, coal and oil). The acid rain affects lakes, rivers and the animals living within them.

Industry is looking for solutions to the air pollution problem. The best solution would be to stop burning fossil fuels all together but at this time we have no other alternatives. One way is to wash the coal before they burn it. Washing removes most of the sulphar but not all of it. Another solution is to clean the smoke before it is released in to the air. Many filters have been installed in smoke stacks but again this solution is not 100 per cent effective.

Another major environmental problem caused by air pollution is global warming or the "green house effect". The green house effect is caused by solar energy being trapped by $CO_{2,}$ Methane, Ozone (O_3) Nitrus gas and Chlorofluoro Carbons (CFCs) that are in the air. The solar energy hits the earth through atmosphere as light, earth lose some of its energy and becomes heat. The earth has a natural greenhouse effect caused by the CO_2 in the atmosphere.

CO_2 is released in to the air from factories and from the exhausts of cars. Some of CO_2 from atmosphere is used by plants for making their food, but forests being cut down it make our problems worse. Cs are released in to the air from aerosol cans. They are also used in air conditioners and refrigerators and leaks from these places. Ozone is a very toxic gas and is good at filtering out UV light in the stratosphere but is bad for us when it is in the lower atmosphere.

Measures

- ☆ The catalytic converter on cars (*i.e.* on exhaust pipe) reduces the amount of CO_2 that is released in to the atmosphere.
- ☆ To control the use of Cs, Aerosol cans are supposed to be banned.
- ☆ Slows the rate of deforestation and plant more trees.

The last major environmental problem caused by air pollution is OZONE depletion. The Ozone is three oxygen molecules (O_3). It is concentrated in the stratosphere about 20-25 km above the earth surface. Ozone protects us from shriveling up under the sun's powerful rays. It blocks out the harmful ultraviolet (UV) light which causes skin cancer. Two major holes have been discovered in the ozone layer, a large one above the South Pole and a smaller one above the North Pole. These two areas have absolutely no protection from UV light.

3.5.2 Water Pollution

Water is our most valuable resource and it is more important than gold. Water is used for irrigation of farm field, cooking, washing clothes, flushing toilets etc and industrial process requires water to function. Only 3 per cent of all water is fresh and drinkable and at that 3 per cent, 75 per cent is frozen, which leaves a grand total of only 1 per cent of the earth's surface water that is readily available for consumption.

Sources of Water Pollution

Organic Pollution

It is becoming more and pressing on the environmental because of the growing population of the world. *e.g.* more many people many people-more waste. Sewage plants do their best, but the secondary discharge that gets in to the water supply causes great problems.

The excess waste acts as a fertilizer or food source for algae and the growth rate is uncontrollable.

- There are a lot of dead algae on the shore of lake and the water is clouded with algae. This situation is known as eutrophication, where algae growth is out of control and the water becomes oxygen depleted, because the dead algae goes to the bottom and uses the oxygen in the deeper water to decompose, but if there is too much dead algae all the oxygen is depleted.
- Wetland means a lowland area such as a marsh or swamp that is saturated with moisture. Wetlands reduce the population of water fouls, fish and animals reduce drastically. Wetlands serve a large filter, water retention and flood prevention areas.
- The water runoff presents a major problem. Any pollutants on ground come in contact with water (contamination of the water supply).

- Large sources of runoff pollutants come from farms and pastures. On the farms all the excess fertilizer, pesticides and herbicides gets in to streams and ends up in the lakes.
- The excess of excretion from cattle *i.e.* urine and manure goes straight to the ground water and right back to the rivers and lakes.
- Toxic waste comes from a great variety of industrial plants.
- Nuclear power plants cause thermal pollution (heated water in to river and lakes).

3.6 Transport Disaster

Road, Rail and Air Accidents

Disaster in transport, involves vehicles and people, several aspects such as vehicle condition, neglect of the traffic rules, alcoholism, rash driving and indiscipline pedestrians cause innumerable accidents every day. The death toll in traffic accidents is rapidly increasing all over India.

Onset Type and Warning System

These are suddenly occurring hazards.

Typical Effects

Causalities

The main damage is the death and injuries that take place in the accident.

Structural Damage

The transport system may get disturbed and vehicles may get damaged.

3.6.1 Road Accidents

Road network is laid for better connectivity and services but we find that the number of accidents is also on the rise. The main causes are violation of the traffic

regulations, speeding, drunk driving and poor maintenance of the vehicles and the roads. All these reasons add to the rising number of accidents and road fatalities.

Measures

- ☆ Drive only when you are competent enough and possess a valid license.
- ☆ These best ways to be safe on the road is to follow lane discipline while driving.
- ☆ Be familiar with all the sign boards, road symbols etc and honor them.
- ☆ Use a helmet while riding two wheelers.
- ☆ Be familiars with road markings.
- ☆ Always maintain your vehicle in good condition.
- ☆ Be careful while driving in the rainy season, driving uphill and driving during night.
- ☆ In case an accident is unavoidable, avoid head on collision.
- ☆ Do not use a harsh horn.
- ☆ Do not use mobile phone while driving.
- ☆ Do not carry more passengers than prescribed by the manufacturer.

3.6.2 Air Accident

Many factors govern the safety of passengers in an aircraft. They include technical problems, fire, landing and takeoff conditions, the environment an airline operates in (mountainous terrain or frequent storm), factors like airport security in cases of hijacking, bombing attempts *etc.*

Air Safety Topics

1. Lightning- Air liners are struck by lightning i.e twice per year.
2. Ice and snow
3. Engine failure
4. Metal fatigue
5. Delamination- risk is as old as composite material (*e.g.* plywood) when subjected to cyclic stress; the fibers may tear off the matrix, the layers of the material then separate from each other a process called Delamination.
6. Stalling- angle of attack.
7. Fire
8. Bird strike
9. Ground damage
10. Volcanic ash
11. Humans factors
12. Terrorism.

Measures

- ☆ Pay attention to the flight crew safety demonstration.
- ☆ Carefully read the safety briefing cards.
- ☆ Know where the nearest emergency exit is and know how to open it in case of emergency.
- ☆ Always keep your seat belts fastened when you are seated.
- ☆ Stay and calm listen to the crew members and do what they say. The cabin crew's most important job is to help you leave safely.
- ☆ Remer smoke rises, follow the track of emergency lights embedded in the floor; they lead to an exit. If you have a cloth, put it over your nose and mouth.

3.6.3 Rail Accident

The most common type of rail accident is derailment due to lack of maintenance, human error or sabotage. Various type of dangerous cargo is also transported such as fuel, oil products, *etc.*

Measures

- ☆ At railway crossings pay attention to the signal and the swing barrier. Do not get underneath and try to get across.
- ☆ In case of unmanned crossing, get down from your vehicle and look at either sides of the track before crossing the track.
- ☆ Do not stop the train on the bridge or tunnel where evacuation is not possible.

- ☆ Do not carry inflammable materials on trains.
- ☆ While on a moving train, do not stand and lean out of the door.
- ☆ Do not smoke in the train.
- ☆ Do not pull the emergency chain unnecessarily.

Chapter 4

Disaster Management

4.1 Disaster Management Cycle

Disaster management is a cyclical process; the end of one phase is the beginning of another (see diagram below), although one phase of the cycle does not necessarily have to be completed in order for the next to take place. Often several phases are taking place concurrently. Timely decision making during each phase results in greater preparedness, better warnings, reduced vulnerability and/or the prevention

of future disasters. The complete disaster management cycle includes the shaping of public policies and plans that either addresses the causes of disasters or mitigates their effects on people, property, and infrastructure. The mitigation and preparedness phases occur as improvements are made in anticipation of an event. By embracing development, a community's ability to mitigate against and prepare for a disaster is improved. As the event unfolds, disaster managers become involved in the immediate response and long-term recovery phases.

The diagram shows the Disaster Management Cycle.

- **Mitigation:** Measures put in place to minimize the results from a disaster. Examples: building codes and zoning; vulnerability analyses; public education.
- **Preparedness:** Planning how to respond. Examples: preparedness plans; emergency exercises/training; warning systems.
- **Response:** Initial actions taken as the event takes place. It involves efforts to minimize the hazards created by a disaster. Examples: evacuation; search and rescue; emergency relief.
- **Recovery:** Returning the community to normal. Ideally, the affected area should be put in a condition equal to or better than it was before the disaster took place. Examples: temporary housing; grants; medical care.

4.2 Disaster Mitigation

Mitigation refers to all actions taken before a disaster to reduce its impacts, including preparedness and long-term risk reduction measures.

4.2.1 Mitigation Categories

Mitigation activities fall broadly into two categories:

Structural Mitigation

Construction projects which reduce economic and social impacts.

Non-Structural Activities

Policies and practices which raise awareness of hazards or encourage developments to reduce the impact of disasters.

Mitigation includes reviewing building codes; vulnerability analysis updates; zoning and land-use management and planning; reviewing of building use regulations and safety codes; and implementing preventative health measures. (World Development Report, 1998) Mitigation can also involve educating businesses and the public on simple measures they can take to reduce loss or injury, for instance fastening bookshelves, water heaters, and filing cabinets to walls to keep them from falling during earthquakes. Ideally, these preventative measures and public education programmes will occur before the disaster. From time to time some mitigation requirements may be outside of the scope of the disaster manager; however, this does not lessen the role to be played by mitigation. On the contrary, it is the responsibility of the emergency manager to avail him or herself with the requisite information to engage community involvement. The primary focus of disaster management is to prevent disasters wherever possible or to mitigate those which are inevitable.

Four sets of tools that could be used to prevent or mitigate disasters include:

1. Hazard management and vulnerability reduction
2. Economic diversification
3. Political intervention and commitment
4. Public awareness

4.2.2 Mitigation Strategies or Measures

- ☆ **Adjusting normal development programmes to reduce losses:** For instance, varieties of crops that are more wind, flood or drought resistant can often be introduced in areas prone to floods, drought and cyclones.
- ☆ **Economic diversification:** In areas where the principal or sole source of the income may be threatened, attempts should be made to diversify the economy and introduce the economic activities that are less vulnerable. Diversification is extremely important where economies are dependent on a single cash crop.
- ☆ **Developing disaster resistant economic activities:** Some economic activities are relatively unaffected by disasters. For instance, situating warehouses in flood plains may be more appropriate than manufacturing plants in the same location. Coconut palms could be more suitable than other fruit trees in cyclone-prone coastal areas. Efforts should be made to identify and encourage the development of enterprises that are less vulnerable to the hazards.

4.2.3 Disaster Preparedness

Disaster preparedness is defined as a continuous and integrated process involving a wide range of activities and resources from multi-sectoral sources. (*Disaster Preparedness Training Programme; International Federation of Red Cross and Red Crescent Societies, IFRCRCS, 2005*). In order that disaster preparedness is undertaken with rewarding outcomes, those involved in the process must approach it from a mitigative, response, recovery and business continuity perspective. That is, when considering disaster preparedness the phases of emergency management must be looked at carefully. Disaster mitigation policies and measures will not stop a disaster especially a natural one from occurring and persisting. What mitigation policies and measures seek to do is reduce vulnerability to, or increase resilience to, the effects of the inevitable disasters to which a country is prone. Basically disaster mitigation and preparedness go hand in hand. Disaster preparedness for example includes implementation of mitigation measures to ensure that existing infrastructure can withstand the forces of disasters or that people can respond in their communities and at the same time protect themselves. The collective capabilities of the country, people, and the government to deal with extreme hazards or adversities when they occur are measures of their cumulative preparedness. In local circumstances and because of historical proneness to disasters, mitigation is important, but preparedness is doubly important. Disaster preparedness involves the preparation of people and essential service providers in their communities for the actions that they will take in case of disasters. If this is the case, consideration must be given to the manner in which the formal responders (Police and Fire Services, Emergency Medical Services personnel and the Military) prepare to respond to disasters. For example, the personnel in these response agencies may have to learn the use of new equipment, treatment methods for diseases or providing services to prevent the escalation of the effects of disasters that will further destroy lives and devastate property.

4.2.4 Disaster Response

The aim of emergency response is to provide immediate assistance to maintain life, improve health, and to support the morale of the affected population. Such assistance may range from providing specific but limited aid, such as assisting refugees with transportation, temporary shelter, and food, to establishing semi-permanent settlement in camps and other locations. It also may involve initial repairs to damaged infrastructure. The focus in the response phase is on meeting the basic needs of the people until more permanent and sustainable solutions can be found. Humanitarian organizations are often strongly present in this phase of the disaster management cycle. During a disaster, humanitarian agencies are often called upon to deal with immediate response and recovery. To be able to respond effectively, these agencies must have experienced leaders, trained personnel, adequate transportation and logistic support, appropriate communications, and guidelines for working in emergencies. If the necessary preparations have not been made, the humanitarian agencies will not be able to meet the immediate needs of the people. This section identifies the principal activities of disaster response. Each activity is (formally or informally) governed by a set of policies and procedures, typically under the auspices of a lead agency. In the end, disaster response activities are implemented by multiple government organizations, international and national agencies, local entities and individuals, each with their roles and responsibilities.

4.2.4.1 Aims of Disaster Response

The overall aims of disaster response are:

- ☆ To ensure the survival of the maximum possible number of victims, keeping them in the best possible health in the circumstances.
- ☆ To re-establish self-sufficiency and essential services as quickly as possible for all population groups, with special attention to those whose needs are greatest: the most vulnerable and underprivileged.
- ☆ To repair or replace damaged infrastructure and regenerate viable economic activities. To do this in a manner that contributes to long term development goals and reduces vulnerability to any future recurrence of potentially damaging hazards.
- ☆ In situations of civil or international conflict, the aim is to protect and assist the civilian population, in close collaboration with the International

Committee of the Red Cross (ICRC) and in compliance with international conventions.

- In cases involving population displacements (due to any type of disaster) the aim is to find durable solutions as quickly as possible, while ensuring protection and assistance as necessary in the meantime.

4.2.4.2 Disaster Response Activities

The following are typical activities of emergency response

1. Warning

Warning refers to information concerning the nature of the danger and imminent disaster threats. Warnings must be rapidly disseminated to government officials, institutions and the population at large in the areas at immediate risk so that appropriate actions may be taken, namely, either to evacuate or secure property and prevent further damage. The warning could be disseminated via radio, television, the written press, telephone system and cell phone.

2. Evacuation and Migration

Evacuation involves the relocation of a population from zones at risk of an imminent disaster to a safer location. The primary concern is the protection of life of the community and immediate treatment of those who may be injured. Evacuation is most commonly associated with tropical storms but is also a frequent requirement with technological or industrial hazards. For evacuation to work there must be:

- A timely and accurate warning system,
- Clear identification of escape routes,
- An established policy that requires everyone to evacuate when an order is given,
- A public education programme to make the community aware of the plan.

In the case of a slow onset of a disaster, for example severe drought, the movement of people from the zone where they are at risk to a safer site is not, in fact, evacuation, but crisis-induced migration. This movement is usually not organized and coordinated by authorities but is a spontaneous response to thc perception by the migrants that food and/or security can be obtained elsewhere.

3. Search and Rescue (SAR)

Search and rescue (SAR) is the process of identifying the location of disaster victims that may be trapped or isolated and bringing them to safety and medical attention. In the aftermath of tropical storms and floods, SAR usually includes locating stranded flood victims, who may be threatened by rising water, and either bringing them to safety or providing them with food and first aid until they can be evacuated or returned to their homes. In the aftermath of earthquakes, SAR normally focuses on locating people who are trapped and/or injured in collapsed buildings.

4. Post-disaster Assessment

The primary objective of assessment is to provide a clear, concise picture of the post-disaster situation, to identify relief needs and to develop strategies for recovery. It determines options for humanitarian assistance, how best to utilize existing resources, or to develop requests for further assistance.

5. Response and Relief

When a disaster has occurred response and relief have to take place immediately; there can be no delays. It is therefore important to have contingency plans in place. Relief is the provision on a humanitarian basis of material aid and emergency medical care necessary to save and preserve human lives. It also enables families to meet their basic needs for medical and health care, shelter, clothing, water, and food (including the means to prepare food). Relief supplies or services are typically provided, free of charge, in the days and weeks immediately following a sudden disaster. In the case of deteriorating slow-onset emergency situations and population displacements (refugees, internally and externally displaced people), emergency relief may be needed for extended periods.

6. Logistics and Supply

The delivery of emergency relief will require logistical facilities and capacity. A well-organized supply service is crucial for handling the procurement or receipt, storage, and dispatch of relief supplies for distribution to disaster victims.

7. Communication and Information Management

All of the above activities are dependent on communication. There are two aspects to communications in disasters. One is the equipment that is essential for information flow, such as radios, telephones and their supporting systems of repeaters, satellites, and transmission lines. The other is information management: the protocol of knowing who communicates what information to whom, what priority is given to it, and how it is disseminated and interpreted.

8. Survivor Response and Coping

In the rush to plan and execute a relief operation it is easy to overlook the real needs and resources of the survivors. The assessment must take into account existing social coping mechanisms that negate the need to bring in outside assistance. On the other hand, disaster survivors may have new and special needs for social services to help adjust to the trauma and disruption caused by the disaster. Participation in the disaster response process by individuals to community organizations is critical to healthy recovery. Through these appropriate coping mechanisms will be most successfully developed.

9. Security

Security is not always a priority issue after a sudden onset of disasters. It is typically handled by civil defence or police departments. However, the protection of the human rights and safety of displaced populations and refugees can be of paramount importance requiring international monitoring.

10. Emergency Operations Management

None of the above activities can be implemented without some degree of emergency operations management. Policies and procedures for management requirements need to be established well in advance of the disaster.

11. Rehabilitation

Rehabilitation consists of actions taken in the aftermath of a disaster to enable basic services to resume functioning, assist victims' self-help efforts to repair dwellings and community facilities, and to facilitate the revival of economic activities (including agriculture). Rehabilitation focuses on enabling the affected populations (families and local communities) to resume more-or-less normal (pre-disaster) patterns of life. It may be considered as a transitional phase between (i) immediate relief and (ii) more major, long-term reconstruction and the pursuit of ongoing development.

12. Reconstruction

Reconstruction is the permanent construction or replacement of severely damaged physical structures, the full restoration of all services and local infrastructure, and the revitalization of the economy (including agriculture). Reconstruction must be fully integrated into ongoing long term development plans, taking account of future disaster risks. It must also consider the possibilities of reducing those risks by the incorporation of appropriate mitigation measures. Damaged structures and services may not necessarily be restored in their previous form or locations. It may include the replacement of any temporary arrangements established as a part of the emergency response or rehabilitation. Under conditions of conflict, however, rehabilitation and reconstruction may not be feasible. For obvious reasons of safety and security, activities in rehabilitation and reconstruction may need to wait until peace allows them.

4.2.4.3 Modern Methods of Disaster Response

New technologies can be very useful and powerful tool in disaster response, namely:

1. **Cell phones:** Cell phones as warning devices can be very useful. Short messages can be sent to recipients warning of imminent threat of tropical storms, wind storms or any severe weather likely to cause damage.
2. **Spatial information:** Use of satellite imagery. The emergency management community is keenly aware of the potential of mapping technologies such as geographic information systems (GIS), remote sensing (satellite imagery), and global positioning systems (GPS) in support of emergency response operations. Increasingly, geographic technologies are being utilized for hazard mitigation as well as response efforts. These range from damage assessments mapping the event and affected areas to search and rescue, risk assessment, risk perception (Hodgson and Palm, 1992), and risk communication (Hodgson and Cutter, 2001).

3. **Social media and social networking:** Social media and social networking can be used as a tool to emergency response communications. Text messaging such as Twitter and the social networking system such as Face book can be used as a channel of communication in disaster response.

4.2.5 Disaster Recovery

As the emergency is brought under control, the affected population is capable of undertaking a growing number of activities aimed at restoring their lives and the infrastructure that supports them. There is no distinct point at which immediate relief changes into recovery and then into long term sustainable development. There will be many opportunities during the recovery period to enhance prevention and increase preparedness, thus reducing vulnerability. Ideally, there should be a smooth transition from recovery to on-going development. Recovery activities continue until all systems return to normal or better. Recovery measures, both short and long term, include returning vital systems to minimum operating standards; temporary housing; public information; health and safety education; reconstruction; counselling programmes; and economic impact studies. Information resources and services include data collection related to rebuilding, and documentation of lessons learned. Additionally, there may be a need to provide food and shelter for those displaced by the disaster.

4.2.5.1 Disaster Response Activity Classification

Recovery activities are classified as short-term and long-term. During response, emergency action was taken to restore vital functions while carrying out protective measures against further damage or injury.

1. **Short-term recovery** is immediate and tends to overlap with response. The authorities restore interrupted utility services, clear roads, and either fix or demolish severely damaged buildings. Additionally, there may be a need to provide food and shelter for those displaced by the disaster. Although called short-term, some of these activities may last for weeks
2. **Long-term recovery** may involve some of the same activities, but it may continue for a number of months, sometimes years, depending on the severity and extent of the damage sustained. For example, it may include the complete redevelopment of damaged areas. The goal is for the community to return to a state that is even better than before the emergency. This is an ideal time to implement new mitigation measures so that the community is better prepared to deal with future threats and does not leave itself vulnerable to the same setbacks as before. Helping the community to take new mitigation steps is one of the most important roles during the recovery phase.

4.2.5.2 The Recovery Plan

The recovery process should be understood clearly and it is important to have a general plan for recovery which should be appended to emergency operation plans. The primary purpose of the plan is to spell out the major steps for managing successful recovery. For each step you will also designate key partners and their roles and steps to mobilize them. The plan should have at least the following seven steps:

1. Gathering basic information
2. Organizing recovery
3. Mobilizing resources for recovery
4. Administering recovery
5. Regulating recovery
6 Coordinating recovery activities
7. Evaluating recovery

For the majority of disasters, local communities are able to provide the assistance needed for recovery. However, for a major disaster, it may be necessary to obtain assistance from the government and other sources. Therefore, preparations must be made to request outside aid if a major disaster occurs. This will mean informing and convincing decision makers, especially those outside the affected area. Documenting the effects of the disaster is the best way to carry this out. Documentation involves providing evidence of what happened. Photographs of the damage provide irrefutable evidence. Take pictures of the damage, the repair work, and completed restorations. You cannot take too many pictures. There can be a good documentation if the following five simple steps are followed:

1. Take pictures of damages and repairs. More is better than too little. Private citizens may have excellent shots to supplement your own.
2. Take notes on damages and repairs. Again, more is better than too little. If there is too much to write at one time, dictate your notes into a tape recorder for later transcription.
3. Clip and file newspaper reports and stories. If you can get video footage from the television stations, do that also.
4. Record all expenditures carefully and keep all receipts and invoices.
5. Make sure anyone acting on behalf of the jurisdiction does the same.

4.2.5.3 Duties of Response Personnel

Each duty involves a series of tasks and steps that must be considered and resolved by decisions and actions. These duties when supported by the response communities are the framework of an appropriate, survival oriented response to hazardous materials incidents. Reasoned decisions based on this approach will minimize the harm resulting from a hazardous material incident and reduce the risk to responders.

4.2.5.4 Community Mitigation Goals

Hazardous materials are usually transported on the roads and railroad networks throughout countries. Some of these hazardous materials are stored and consumed by the community, in particular, gasoline for vehicles, propane for heating, and anhydrous ammonia for fertilizers. During disasters, the potential is high for release or spill of hazardous material into residential areas or areas frequented by communities. It is important therefore to raise and promote awareness of communities in the safeguard, handling, use and disposal of hazardous materials.

4.2.5.5 Pre-disaster Mitigation Plan

To increase the public's awareness of the full range of man-made or technological hazards, it is recommended that education and outreach programmes are developed and implemented. Actions to take include:

1. Educating the public about the hazardous materials to which they are most frequently exposed.
2. Help homeowners identify Hazardous Materials from which they are at risk.
3. Identify, publish and disseminate a procedures manual on the disposal of hazardous materials.

4.2.5.6 Personnel Training

Training personnel is the preparation of resource people to provide basic information on appropriate targeted goals. It provides premier world-class training, products and services through innovative methods and technologies that contribute to the protection of life and property in the environment. It is a training that develops resources based on the needs of people.

4.2.5.7 Purpose

The reason for disseminating quality information in the informal mode is very important to the communities. The obvious reason is to integrate the local skills and knowledge with modern technologies with the immediate resources that are available especially with regards to disaster risk management. The resource people in the government, NGOs and communities within the local government structure are driving the personnel training programmes.

4.2.5.8 Volunteer Assistance

A volunteer assistance is a group of people or organizations that are willing to give assistance on voluntary basis. This group of people or organization provide predictable, safe and sanitary environment in the aftermath of a disaster in the communities. They participate in the community organs and provide and liaise relief from the wider community. For example, the cultural values play an important role in assisting families receives materials and utensils if they have lost most or all of their well-being. The next level where volunteer assistance comes from is organizations from outside but within the national boundaries.

4.2.5.9 School-based Programmes

School-based programmes on disaster management are sets of activities that are dealing with disaster prevention strategies for the school. The development of all school-based programmes on disaster management should begin with a determination of which natural and technological disasters are possible in the school area. Make sure all the school communities do not assume they know all the disaster risks. School stakeholders may be surprised to learn that their school areas are subject to natural disasters they had not anticipated. Also, remember that disasters can have a cascading effect. For example, think about how transportation routes or other external factors may also affect the schools by asking "Are we near a major highway where hazardous chemicals are transported, putting our school in danger of a chemical spill?" Once schools find out what disasters are possible in their areas, assess their structures. For example, falling objects, fires, and the release of hazardous materials, flying debris and roof collapse, cause most of the injuries and deaths related to disasters. Be sure, then, to look for such hazards when doing their assessment. School-based programmes on disaster management consist of conducting survey in a systematic manner, making an inventory of all items that require attention. It may be possible to enlist volunteers from among their parents or emergency management community. Since prevention of disaster risks is everybody business, all school stakeholders must personally walk the school halls and classrooms to determine what risks exist. Before a disaster, schools should document their property, something that can be done as part of the hazard assessment. Schools that take photos and videos prior are far ahead in recovery with less hassle and more quickly restored than the schools where files are missing and records were not kept.

4.3 Emergency Management Systems

4.3.1 What is Emergency Management?

Emergency management is a discipline that involves the avoidance of risks, while simultaneously putting plans in place to deal with disasters and emergency situations if and when they do occur with a view to rebuild and restore society to a functional level in as short a time as possible after a disaster. Emergency management is therefore a shared responsibility between government and citizens of a country towards building a sustainable, disaster-resilient society. The ultimate purpose of emergency management is to:

- ✰ Save lives
- ✰ Preserve the environment
- ✰ Protect property
- ✰ Protect the economy

4.3.2 What are Emergency Management Systems (EMS)?

Emergency management systems are technological aids that facilitate the effective management of disasters. EMS technology can assist in several areas that are critical to effective disaster management, such as:

- ✰ Drafting and testing of evacuation and general disaster plans (*Evacuation Plans*).
- ✰ Establishment of shelters as well as informing the public of shelter locations, items that should be taken to the shelter and general "shelter behaviour".

- Training personnel in effective shelter management, basic first aid and other "response" skills (*Manpower*).
- Establish a national warehouse and ensure that it is stocked with items for national survival in the immediate aftermath of the disaster, before the arrival of overseas help (*Materials*).
- Setting-up reliable communication systems, such as, the traditional two-way CB-type radios (*Communication*).
- Putting transportation plans in place, which should include air transportation to facilitate air-lifts and rescues, delivery of food supplies to severely affected areas cut-off from vehicular traffic and comprehensive damage assessment activity (*Transportation*).

Chapter 5

Disaster Management and GIS-GPS-RS

5.1 Disaster Management and GIS

5.1.1 What does GIS Mean?

Geographic Information Systems are information systems capable of integrating, storing, editing, analyzing, sharing, and displaying geographically-referenced information. In a more generic sense, GIS is a tool that allows users to create interactive queries (user created searches), analyze the spatial information, edit data, maps, and present the results of all these operations.

5.1.2 GIS Applications

GIS applications can be useful in the following activities:

1. **To create hazard inventory maps:** At this level GIS can be used for the pre-feasibility study of developmental projects, at all inter-municipal or district level.
2. **Locate critical facilities:** The GIS system is quite useful in providing information on the physical location of shelters, drains and other physical facilities. The use of GIS for disaster management is intended for planners in the early phase of regional development projects or large engineering projects. It is used to investigate where hazards can be a constraint on the development of rural, urban or infrastructural projects.
3. **Create and manage associated database:** The use of GIS at this level is intended for planners to formulate projects at feasibility levels, but it is

also used to generate hazard and risk maps for existing settlements and cities, and in the planning of disaster preparedness and disaster relief activities.

4. **Vulnerability assessment:** GIS can provide useful information to boost disaster awareness with government and the public, so that (on a national level) decisions can be taken to establish or expand disaster management organisations. At such a general level, the objective is to give an inventory of disasters and simultaneously identify "high-risk" or vulnerable areas within the country.

5.1.3 GIS and the Disaster Management Cycle

5.1.3.1 Planning

The most critical stage of disaster management is the realization that there is a need for planning based on the risk that is present. The extent to which lives and properties will be spared the adverse effects of a disaster is dependent on the level of planning that takes place and the extent to which technology has been incorporated in planning efforts. GIS is useful in helping with forward planning. It provides the framework for planners and disaster managers to view spatial data by way of computer based maps.

5.1.3.2 Mitigation

The use of GIS in disaster management can help with structural and nonstructural mitigation. GIS allows you to spatially represent areas at risk and the level of risk associated with a particular hazard, which can be a guide in decision making. It will facilitate the implementation of necessary mechanisms to lessen the impact of a potential emergency. With GIS, disaster managers are in a better position to determine the level of mitigative structures that should be in place given the vulnerability of an area or population.

5.1.3.3 Preparedness

As a tool, GIS can help with the identification and location of resources and "at risk" areas. It establishes a link between partners and critical agencies, which allow disaster managers to know where relevant partner agencies are stationed. In the context of disaster management, GIS maps can provide information on the human resources present in an Emergency Operation Centre as well as on the ground personnel such as security, health providers and other key responders. This is particularly useful since the technology can help with strategic placement of emergency personnel where it matters most. GIS helps to answer the question of who is to be based where and at what phase during the emergency. It can help to determine whether or not road infrastructure and communications networks are capable of handling the effects of disaster and, if necessary, guide in the placement of resources.

5.1.3.4 Response

GIS technology can provide the user with accurate information on the exact location of an emergency situation. This would prove useful as less time is spent trying to determine where the trouble areas are. Ideally, GIS technology can help to provide quick response to an affected area once issues (such as routes to the area) are known. In the case of a chlorine explosion for example, GIS can indicate the unsafe area as well as point rescue workers to resources that are closest to the affected areas. GIS can be used as a floor guide for emergency response to point out evacuation routes, assembly points and other evacuation matters.

5.1.3.5 Recovery

Mapping and geo-spatial data will provide a comprehensive display on the level of damage or disruption that was sustained as a result of the emergency. GIS can provide a synopsis of what has been damaged, where, and the number of persons or institutions that were affected. This kind of information is quite useful to the recovery process.

5.1.4 Advantages of GIS

GIS as an innovative and interactive technology tool has more advantages than there are challenges.

1. GIS has the ability to represent spatial information over a wide geographic area. GIS accommodates 3-dimensional graphics which will provide a more detailed viewed of its contents.
2. GIS technology facilitates the integration of different geo-spatial information; which can include models, maps and other graphic forms.
3. GIS effectively analyzes, collects, manages and distributes up-to-date information.
4. GIS is versatile and easy to use – this requires little training to get individuals involved in the process.
5. Attribute table which forms a database- Given that information from GIS can be easily tabulated; it provides a comprehensive pictorial overview of what is happening in the country. For example, GIS can show the exact location of shelters across the country, or the sites where search and rescue operations have taken place.

5.1.5 Challenges of using GIS in Disaster management

1. Major impacts on life of people, economy and environment. In the context of emergency management, GIS can impact people's lives in a significant way as it reveals sometimes personal and people specific information.
2. Crucial decisions- Based on the information obtained from GIS mapping, it may require taking critical (sometimes hard) decisions in the best interest of the affected area.
3. GIS being a technological tool can be complex and a bit difficult to grasp initially.
4. Large amounts of information (input) is usually required to get useful output from the system.

5. Time is critical during an Emergency- The decision-making process may be stalled during an emergency due to:
   - the large volume of information required by the GIS system; and
   - the vast amount of time require to analyze the information before a decision is finally made.

5.1.6 Who can Use GIS ?

GIS can be used in any area of disaster management. Among the professionals within the disaster management discipline who would find GIS useful are:

- Emergency Planners
- Meteorologists
- Geologists
- Telecommunications personnel
- Security personnel
- Health practitioners

5.2 Global Positioning System (GPS) and Disaster Management

5.2.1 What is GPS?

The term Global Positioning System (GPS) is used to refer to the Global Navigation Satellite System (GNSS) developed by the United States Department of Defense. The proper name is The Navigation System with Timing and Ranging Global Positioning System (NAVSTAR GPS) however the acronym GPS is typically used. Though initially intended solely for US military purposes the GPS system was extended for civilian use in the 1980's. Popular applications include automobile and marine navigation, tracking, farming and research. GPS is a grouping of 24 well-spaced satellites that orbit the earth and make it possible for people with ground receivers to pin-point their exact geographic location with great accuracy. GPS equipment is widely used across the globe and is sufficiently "low-cost" so that anyone can own a GPS receiver.

5.2.2 Application of GPS to Disaster Management

GPS is particularly useful during disasters because it operates in any weather, anywhere and at all times. While it functions simply to give the location of the receiver, the level of precision of GPS makes it quite useful in disaster management. In many instances GPS data is integrated with GIS to overlay real-time activity during emergency. GPS find its greatest utility during the response and recovery phases; however it can also be utilized during preparedness and mitigation phases. An important application of GPS in EDM is tracking of emergency vehicles or supplies. In this application the GPS receiver attached to the vehicle and the location is overlaid onto a map. Other applications include the monitoring the height of waves. GPS units are fixed to buoys and the height of the units are can be determined to within centimeters any significant change in wave height or velocity can trigger an alarm for a tsunami or sea surge. Volcanoes can also be monitored using GPS. By measuring the deformation of the ground, inferences about volcanic activity can be made.

5.3 Remote Sensing and Disaster Management

5.3.1 What is Remote Sensing?

Remote sensing is the use of electromagnetic (EM) wave radiation to acquire information about an object or phenomenon, by a recording device that is not in physical or intimate contact with the object. In other words, Remote Sensing is the acquisition of information about an object by a recording device that is NOT in physical or intimate contact with the object. As you read this material you are actually engaging in remote sensing; we do this so naturally that we seldom realize it. We could take this a step further - we use telescopes to view distant planets. We are definitely sensing objects remotely. In both cases the sensor is our eyes and the EM wave is light. If the term EM waves seems new to you it shouldn't. Everyday light, radio waves and microwaves and X-rays are examples of EM waves. EM waves transport energy and information from one place to another. They are used in cellular networks, microwave ovens, portable radios, X-ray machines and satellites systems. Remote sensing in the context of disaster management usually refers to the technology that includes man-made sensors that are attached to aircrafts, or satellites. Instead of viewing a far away planet from earth, the sensing equipment is usually high above looking down at our 'distant' planet - earth. Distant in this context can mean just a few hundred feet overhead or miles above the earth's surface.

5.3.2 Advantages of Remote Sensing

- Saves time.
- Users of the technology do not have to be in direct contact with danger zones.
- Shows image of very large areas of land or space.
- Detect features at wavelengths not visible to the human eye.
- Data can be regularly and routinely acquired and archived.
- The most cost-effective dataset for monitoring change over large areas.
- Can assist with damage assessment monitoring.
- The imagery obtained, using remote sensing, can be useful for forward planning and reconstruction of an affected area.
- Helps to prevent the recurrence of the same disaster in the future.

5.3.3 Challenges Faced Using Remote Sensing

- It can be costly to build and operate a remote sensing system.
- Small size activities cannot be delineated on remote sensing imagery or through aerial photography.
- Data can be difficult to interpret and may require expert skills.
- Resolution is often coarse.

Chapter 6

Phases and Activities of Disaster Management

Efforts are to be made to mitigate natural disasters at National and global levels. There are various phases and personal activities related to disaster management:

6.1 Mitigation Phase

It is the phase where measures are taken to reduce loss of life, livelihood and property by disasters, either by reducing vulnerability or by modifying the hazard, where possible. Examples are:

- ☆ Adoption and enforcement of building codes.
- ☆ Utilization of design and construction techniques that will make critical facilities adequacy resistant to damage by hazards.
- ☆ Mitigation involves structural and non structural measures taken to limit the impact of disasters.
- ☆ Structural mitigation: This involves proper layout of building, particularly to make it resistant to disaster.
- ☆ Non structural mitigation: This involves measures taken other than improving to structure of building.

6.2 Preparedness Phase

Preparedness is aimed at preventing a disaster from occurring, personal preparedness focuses on preparing equipment and procedures for use when a disaster occurs *i.e.* planning. Preparedness measures can take many forms indicating the construction of shelters, installation of warning devices, creation of back up life line services (*e.g.*, power, water, and sewages) and rehearsing evacuation plans. The preparation of survival kit such as a "72- hour kit" is often advocated by authorities. These kits may include food, medicine, flashlights, candles and money.

6.3 Response Phase

Action carried out in a disaster situation with the objective to save lives, alleviate suffering and reduce economic losses. Actions include search and rescue and the provision of shelter, water, food and medical care.

6.4 Recovery Phase (Reconstruction Phase)

The recovery phase starts after the immediate threat to human life has subsided. During reconstruction, medium and long term repair of physical, social and economic damage and return of affected structures to condition equal to or better than before the disaster.

6.5 Concepts Relevant to Disaster Management

- ☆ **Disaster:** A disaster is a serious disruption of the functioning of a society, causing widespread loss in: 1) Life, 2) Property and 3) Environment.
- ☆ **Hazard:** Hazard is defined as the probability of the occurrence of dangerous phenomenon at a given place within a given period of time. A hazard becomes a disaster when it results in a sufficiently large scale loss of life and property.
- ☆ **Vulnerability:** Vulnerability is a set of prevailing or consequential conditions which adversely affect the ability to prevent mitigate, prepare for respond to an event.

Vulnerability has been divided in to three main categories.

1. Physical/material vulnerability
2. Social/organizational vulnerability and
3. Attitudinal/motivational vulnerability.

- ☆ **Risk:** Risk is defined as the product of hazard and vulnerability statistical probability damage to particular element which is at risk, from a particular source or origin of hazard (Lewis, 1999).
- ☆ **Disaster risk:** Disaster risk, per se, incorporates hazard (a source of danger) multiplied by value (anything of human value including life, property and/or livelihood), multiplied by vulnerability (extent of exposure to hazard impact) (van open, 2001). Disaster risk can be reduced by any technique that reduces the impact of the hazardous event.
- ☆ **Mitigation:** The measures which can be taken to minimize the destructive and disruptive effects of hazards and thus lessen the magnitude of a disaster.
- ☆ **Prevention:** This covers activities designed to impede the occurrence and/or effect of a disaster.

- **Emergency response:** It includes essential services and activities that are undertaken immediately after the occurrence of a disaster to assist disaster victims.
- **Trigger mechanism:** The trigger mechanism envisages that on receiving signals of a disaster happening or likely to happen, all the resources and activities required for the mitigation process are energized, and activated simultaneously without any loss of time, and the management of the event is visible on the ground.
- **Prediction:** It involves collecting the necessary information to determine the possibility of a hazard so as to initiate the preventive andmitigation activities.
- **Monitoring:** It implies tracking or recording a disaster as it happen to provide timely information to help the affected persons and those responsible for performing the relief and recovery activities.
- **Assessment:** Assessment is the post event activity of collecting information and data that can be used to estimate the extent of damage caused by a disaster. Different dimensions of damage like: loss of property, loss of productivity capacity, monetary loss, *etc.* and their impact are generally estimated.
- **Disaster management:** It is basically a management intervention for reducing the negative impacts of disaster on life, property and ecosystem, building preparedness for accepting and coping up with the challenges and risks generated from disaster.
- **Disaster preparedness:** It refers to a specialized type of management strategy for unavoidable circumstances (such as flood, drought *etc.*) where the communities or systems are made prepared not to succumb at the disaster consequences.
- **Rehabilitation:** Rehabilitation mainly refers addresses sustainable reconstruction and income generating activities of the people, after a disaster has happened.

6.6 Trigger Mechanism

Trigger Mechanism is a quick response mechanism, which would spontaneously set the vehicle of management into motion on the road to disaster mitigation process. The trigger mechanism has been envisaged as a preparedness plan whereby the receipt of a single of an impending disaster would simultaneously energies and activates the mechanism for response and mitigation without loss of crucial time. This would entail all the participating managers to know in advance the task assigned to them and the manner of response. Identification of available resources, including manpower, material and equipment and adequate delegation of financial and administrative powers are prerequisites to successful operation of the trigger mechanism. As and when a disaster takes place, be it natural or man-made, the managers struggle to mitigate its effects on human lives and material losses. The immediate response in all disasters has more or less the same parameters. These are to provide rescue and relief and save the precious human life. Thus, the emergency response of the disaster managers is a factor independent of the types of intensity of the disasters. As and when the disasters strike or take place, the managers are required to swing in action without losing time. Generally, in such situations, the managers start organizing, planning and activating the mitigation process. On the other hand, the event had already taken place and the need of that hour is to start the mitigation process and virtually no time can be spared at that stage for the activities like organizing and planning. Time is the essence of the immediate relief and rescue operations to save human lives and mitigate human miseries for the next 48 to 72 hours. Thereafter, actually what is required to be done is a part of long term rehabilitation and reconstruction programmes.

The trigger mechanism in fact is a preparedness plan in which all the participating managers, and actors know in advance the task assigned to them and the manner in which they have to be prepared themselves to respond. In fact the trigger mechanism is in essence the Standard Operating Procedure (SOP) in which the implementation of the efforts on ground is well laid down. Generally, the activities which include evacuation, search and rescue, temporary shelter, food, drinking water, clothing, health and sanitation, communications, accessibility, and public information which are very important components of disaster management, would follow on the activation of the Trigger Mechanism. All these major activities which are common in all types of disasters will require sub-division and preparation of sub-action plans by each specified authority. They will be required to list all requirements and their availability within the prescribed response time. Separate SOPs need to be in place for each front line agency like Police, Fire- Service, PWD, Highways, Health Departments.

The Trigger Mechanism requires the disaster managers to:

- ✰ Evolve an effective signal/warning mechanism.
- ✰ Identify activities and their levels.
- ✰ Identify sub-activities under each activity/level of activity.
- ✰ Specify authorities for each level of activity and sub-activity.
- ✰ Determine the response time for each activity.
- ✰ Work out individual plans of each specified authority to achieve the activation as per the response time.
- ✰ Have Quick Response Teams for each specified authority.
- ✰ Have alternative plans and contingency measures.
- ✰ Provide appropriate administrative and financial delegations to make the response mechanism functionally viable.
- ✰ Undergo preparedness drills.

6.7 Paradigm Shift towards Prevention and Reduction

Recognizing the rapidly rising world-wide toll of human and economic losses due to natural disasters, the UN General Assembly in 1989 took a decision to launch a far reaching global undertaking during the nineties to save human lives and reduce the impact of natural disasters. With this aim in mind, the decade1990-2000 was declared as the International Decade for Natural Disaster Reduction (IDNDR). The objective of the IDNDR was to reduce, through concerted international action, especially in developing countries, the loss of life, property damage and social and economic disruption caused by natural disasters such as earthquakes, floods, cyclones, landslides, locust infestations, drought and desertification and other calamities of natural origin.

By the year 2000, as per the plan of the IDNDR, all countries should have had:

1. Comprehensive national assessments of risks from natural hazards, with these assessments taking into account their impact on developmental plans,
2. Mitigation plans at national and/or local levels, involving long term prevention and preparedness and community awareness, and
3. Ready access to global, regional, national and local warning systems and widespread dissemination of such warnings.

6.8 Prevention, Mitigation and Preparedness Strategy

- ☆ Development of a culture of prevention as an essential component of an integrated approach to disaster reduction.
- ☆ Prepare and maintain in a state of readiness 'Preparedness and Response Plans' at National, State and District levels.
- ☆ Adoption of a policy of self reliance in each vulnerable area.
- ☆ Education and training in disaster prevention, mitigation and preparedness for enhancement of capabilities at all levels.
- ☆ Identification and strengthening of existing centers of excellence.
- ☆ Order to improve disaster prevention, reduction and mitigation capabilities.

6.9 Ushering in a New Culture of Disaster Management

Culture of Preparedness

Hitherto, the approach towards coping with the effects of natural disasters has been post-disaster management involving many problems such as law and order, evacuation and warnings, communications, search and rescue, firefighting, medical and psychiatric assistance, provision of relief and sheltering, *etc.* After the initial trauma of the occurrence of the natural disaster is over within the first few days or weeks, the phase of reconstruction and economic, social and psychological rehabilitation is taken up by the people themselves and by the government authorities. Soon thereafter the occurrence of the disaster is relegated to historic memory till the next one occurs either in the same area or in some other part of the country. It is not possible to do away with the devastation of natural hazards completely. However, experience has shown that destruction from natural hazards can be minimized by the presence of a well functioning warning system, combined with preparedness on the part of the vulnerable community. Warning systems and preparedness measures reduce and modify the scale of disasters. A community that is prepared to face disasters receives and understands warnings of impending hazards and has taken precautionary and mitigatory measures will be able to cope better and resume their normal life sooner. Culture of Prevention- One of the many lessons learnt by victims of various natural disasters is that the aftermath of a disaster can be even worse than the disaster event itself. Thus, there is a need to acknowledge the necessity for efforts towards disaster prevention. However, people are often surprised by the concept of reducing disasters. How, it is often asked, can a natural disaster such as an earthquake or a cyclone be reduced or prevented? Natural occurrences such as floods, earthquakes, cyclones, *etc.*, simply cannot be avoided altogether, they are a part of the environment we live in. What can be done, however, is to take preventive measures at various levels of society in order to make the impact of such natural hazards as harmless as possible for people and people's properties. The impact of a natural hazard can be reduced; its worst effects can be prevented.

Early Warning

Building codes do not exist against storm surge inundation. Prescribed means today to save life and properties against storm surge inundation is to evacuate people to safer places as quickly as possible on receipt of warnings. Coordinated early warning systems against tropical cyclone are now in existence around the globe and it is possible to warn the affected population at least 24 to 36 hours in advance about the danger from a tropical cyclone. By taking advantage of early warning systems, it is now possible by prepared and knowledgeable communities to minimize the loss of lives and properties.

Development Planning

There is a need to integrate development plans and regulations with disaster-mitigation. The construction of roads, railways lines, bridges, *etc.*, should be according to the topography and geology of that area in terms of risk and vulnerability. All development projects (engineering and non engineering) including irrigation and industrial projects should be targeted towards disaster-mitigation. Environmental protection, afforestation programmes, pollution control, construction of earthquake-resistance structures should have priority for implementation. What is important is to introduce a culture of prevention in disaster managers and all communities, at all levels: action to save lives must be taken before disaster strikes. For instance, most of the deaths and casualties in an earthquake are caused not by the earthquake itself but due to the collapse of buildings and concrete structures. Hence earthquake proof features need to be planned and incorporated at the structural design itself. Retrofitting of existing structures will also mitigate the effects of an earthquake. Such preventive measures are essential also in a State like Tamil Nadu considering that much of the State has been upgraded to Zone III in the revised seismic zonation map of India, on par with Latur in Maharashtra. The building control regulations need to be revised accordingly.

Chapter 7

National Disaster Management (India)

7.1 Search and Rescue Team (SAR): (District Level)

The first few hours after a disaster are known as 'Golden hours' in the parlance of disaster management. Maximum lives can be saved during this period. Therefore, Search and Rescue is considered as the first and the most important task in the past disaster period. A well trained and properly equipped professional search and rescue (SAR) team, at least in each district is therefore considered as highly essential. Acknowledging the importance of SAR teams at the district level, the Government of Maharashtra issued Government. Resolution (GR) in June 2005, for the constitution of Search and Rescue (SAR) team in every district of the state. The 36 members SAR teams have been constituted as per the guidelines of Ministry of Home Affairs, Government. of India. The composition of this SAR team is as follows.

- ☆ Team Commander-1
- ☆ Deputy Team Commander-1
- ☆ Operation Groups-(3-Groups)
- ☆ Group –I: 8 members, Group-II: 8 members.
- ☆ Group- III: 8 members.
- ☆ Technical supports.
- ☆ IT-1, Communication Expert- 1
- ☆ Engineers-2, Doctor-1
- ☆ Paramedic (male)-1, Paramedic (female)-1

- ☆ Administrative support.
- ☆ Administrative officer-1,Suppoprt staff-2

Members of this team have be drawn from Police, fire services, Home Guards and Civil Defence, NGOs, *etc.* Training to all districts team is being imparted by the National Disaster Response force, Pune (M.S.) Also the trained members are doubling as trainers to SAR teams at Taluka and Village level.

7.2 Role of NGOs (Non-Government Voluntary Organizations (Role of NGOs in Disaster Risk Reduction)

Many of the non- government organizations (NGOs) in India have come to play a very useful role in disaster risk reduction. They operate at the grass roots along with the community. The various functions/activities performed by NGOs are enumerated as follows.

7.2.1 Pre Disaster

1. Awareness and information campaigns.
2. Training of local volunteers
3. Advocacy and planning.

7.2.2 During Disaster

1. Immediate rescue and first aid, psychological help and counseling
2. Supply of food, water, medicine and other materials.
3. Ensuring sanitation and hygiene.
4. Damage assessment.

7.2.3 Past Disaster

1. Reconstruction aid *i.e.* Technical and material.
2. Assistance in seeking financial aid.
3. Monitoring.

Role of NGOs is as one of the most effective alternative means of achieving an efficient communication link between the disaster management agencies and the affected community. The roll of NGOs is a key element in disaster management. There are different types of voluntary organizations functioning in India, such as international organization, private volunteer grassroots (local) organizations, besides these, there are interest groups such as Rotary club, Lion's club, Indian Red cross.

There are NGOs who enjoy international support and can respond quickly with large amount of supplies and services.

7.3 Administrative Response Mechanism

On 23 December, 2005, the Government of India (GoI) took a defining step by enacting the Disaster Management Act, 2005, (hereinafter referred to as the Act) which envisaged the creation of the National Disaster Management Authority (NDMA), headed by the Prime Minister, State Disaster Management Authorities (SDMAs) headed by the Chief Ministers, and District Disaster Management Authorities (DDMAs) headed by the Collector or District Magistrate or Deputy Commissioner as the case may be, to spearhead and adopt a holistic and integrated approach to DM. There will be a paradigm shift, from the erstwhile relief-centric response to a proactive prevention, mitigation and preparedness-driven approach for conserving developmental gains and also to minimize losses of life, livelihoods and property.

7.4 National Disaster Management

In India, disaster management is the responsibility of State Governments. Research, survey, guidelines and provisional of financial assistance to the States are the responsibility of the Central Government. The Department of Agriculture and Co-operation (DAC) in the Agriculture Ministry is the nodal department for all matters concerning natural disasters relief at the centre. In the DAC, Relief Commissioner functions as the nodal officer to co-ordinate relief operations for all natural disasters. The Central Relief Commissioner receives information relating to forecast warning of the Natural Disaster/Calamity from the Director General, IMD or from the Central Water Commissioner on a continuity basis. He provides feedback through Agriculture Secretary to the Agril Minister, Prime Minister and the Cabinet. There is a Crises Management Group (CMG) headed by the Cabinet Secretary/Central Relief Commissioner and consists of Nodal Ministries in-charge of various types of disasters. A multi-disciplinary Central Government Team conducts a disaster assessment and also makes recommendation for assistance of affected state. More recently, a National Committee on Disaster Management under the Chairmanship of the Prime Minister has been set up to look in to issues of disaster management planning at National, State and District levels. The CMG meets every six months.

The State also has a State Crises Management Group (SCMG) which functions under the Chairmanship of the Chief Secretary/Relief Commissioner. The group comprises senior officers from the departments of revenue, relief, home, civil supplies, power, irrigation, water supply, panchayat, agriculture, forests, rural development, health, planning, public work and finance. The SCMG is expected to work as per guidance received from the Government. of India from time to time and formulate action plans accordingly for dealing the different natural disaster. The Chief Secretary is the head of State Administration.

7.5 Organizational Structure and Institutional Arrangements for Emergency Management in the State

7.5.1 State Emergency Management Planning Committee (SEMPC)

A State Emergency Management Plan must be prepared for each kind of disaster and the details of the organizational structure for emergency management activities should be made known. Responsibility of concerned agencies for the execution of rescue, relief and recovery operations and Standard Operating Procedure for each should be made available. A State Emergency Management Planning Committee (SEMPC) should be constituted with all the stakeholders as members.

7.5.2 State Crisis Group (SCG)

The setting up of a State Crisis Group (SCG) will enable quick decision making, operational direction and coordination of the issue of warning and execution of rescue, relief and recovery operations. The responsibilities of the SCG would include:

1. On spot decision making.
2. Control and coordination of response and recovery activities.
3. Resource mobilization and replenishment.
4. Monitoring of overall response and recovery activities.
5. Preparation of reports for submission to State Government through.
6. Relief Commissioner.

7.5.3 State Emergency Manager (SEM)

Different government agencies and the NGOs are supposed to operate within the overall direction and coordination of the Commissioner of Revenue Administration/ Relief Commissioner, who may be designated as the State Emergency Manager (SEM). The individual government agencies and the NGOs will perform the assigned jobs but the State Emergency Manager will appropriately augment their resources by drawing upon resources from other government agencies and the local communities. The Chief Secretary or Commissioner of Revenue Administration can also assign additional responsibilities and functions to different Government agencies to meet the requirements of the situation. During the time of emergency the Chief Secretary or the Commissioner of Revenue Administration who is the State Emergency Manager (SEM) would act as the focal point for control and coordination of all activities. His responsibilities would be:

1. Get in touch with the local Army/Navy/Air force units for assistance in rescue, evacuation and relief.
2. Requisition resources, materials and equipment from all Departments/ Organizations of the government and also from private sector.
3. Direct industry to activate their onsite and off -site disaster management plan.
4. Set up Site Operations Centre in the affected area with desk arrangements.
5. Arrange establishment of transit and/or relief camps, feeding centers and cattle camps.
6. Send Preliminary Information Report and Action Taken Report to the Government.
7. Arrange immediate evacuation whenever necessary.

7.5.4 State Emergency Control Room (SECR)

The need for directing the operations at the affected site, the need for coordination at the district headquarters and the need for interaction with the State Government to meet the conflicting demands at the time of disaster is the responsibility of the Relief Commissioner and his team. A well equipped State Emergency Control Room (SECR) in terms of manpower and equipment should be established to help the Relief Commissioner and his team and to perform the following functions:

- ☆ Collection and compilation of information from the affected area
- ☆ Documenting information flow
- ☆ Decision making regarding resource management
- ☆ Allocation of task to different resource organization
- ☆ Supply of information to State Government

The SECR may have senior representatives in the capacity of Desk Officers from the following key resource agencies:

- ☆ Search, Rescue and Evacuation desk - Police and Fire Services
- ☆ Logistics and Welfare desk
- ☆ Medical desk
- ☆ Infrastructure desk

The Desk Officers should maintain constant contact with the State Crisis Group members and the other district heads to ensure quick decision- making.

7.5.5 Activities of State Emergency Control Room (SECR)

(a) Normal Times

The responsibilities during the normal times will include:

- ☆ Ensure all warning and communication systems, instruments are in working condition.
- ☆ Collect information on a routine basis from the State departments on the vulnerability of areas to disasters.
- ☆ Liaise with SEMPC
- ☆ Develop status reports of preparedness and mitigation activities in the State.
- ☆ Ensure appropriate implementation of State Emergency Management Plan.
- ☆ Maintain data bank with regular updating Evaluation and updation of State Emergency Management Plan is the responsibility of SEMPC. However, SEMPC would keep an account of the amendments and accordingly review its response strategy. SECR will be responsible for activating the trigger mechanism in the event of receipt of a warning or occurrence of a disaster.

(b) Activities on Occurrence of Emergency Issue Warning/Alert

On the basis of message received from the forecasting agencies, warning has to be issued for the general public and the departments which play a vital role during emergencies. Issue of correct and timely warning would be one of the prime responsibilities of SECR. For effective dissemination of warning SECR should have a well-planned line of communication. The Relief Commissioner would be the authoritative body to issue warning. Formulation of warning message should consider the target group for which it is issued.

(c) Post-emergency Activities After an Emergency the main Responsibility of a SECR would be:

Evaluation of relief and rehabilitation activities in order to assess the nature of state intervention and support, suitability of the organization structure, institutional arrangements, adequacy of Operating Procedures, monitoring mechanisms, information tools, equipment and communication system.

Post -emergency impact studies for long term preventive and mitigation efforts to be taken.

Communication Room (Main Message Room)

The police wireless system should continue to be in contact with the SECR. In every district the police have a well-established wireless communication system; therefore, under any emergency the communication resources available with the police may be utilized. During disaster, SECR would be connected to Site Operations Centre and the facilities at various Desks. Emergency Support Functions (ESFs) are how Emergency Management accomplishes many of the tasks of responding to an emergency.

7.5.6 List of Emergency Support Functions

ESF No. 1 – Communication

ESF No. 2 – Public Health and Sanitation

ESF No. 3 – Power

ESF No. 4 – Transport

ESF No. 5 – Donation

ESF No. 6 – Search and Rescue

ESF No. 7 – Public Works and Engineering

ESF No. 8 – Food

ESF No. 9 – Information and Planning

ESF No. 10 – Relief Supplies

ESF No. 11 – Drinking Water

ESF No. 12 – Shelter

ESF No. 13 – Media

ESF No. 14 – Help lines

7.5.7 Planning Process

The Planning process for disaster management is based on the principle that response and level of preparedness required are dependent on the extent of vulnerability and the level of capacity to deal with situations. Disasters may be graded at three levels:

L1: District Level Disaster within the capabilities of the district administration to deal with

L2: State Level Disaster within the capabilities of State Government to deal with

L3: National Level Disaster requiring major intervention of the Central Government

L0: No disaster situation- This is the level at which surveillance, preparedness and mitigation activities must be focused on.

7.6 District Level

The District Level Co-ordination and Review Committee is constituted, which is headed by the Collector as the chair person with participation of all other related agencies and departments.

The various measures undertaken at district level are:

1. **Contingency plan**- The District Magistrate develops a district level plan, which is submitted to the State for approval. The plan assigns measures to be taken by the different district departments and their functionaries and it identifies the area of coordination.
2. **District relief committee**- It includes official, non official members and members from local, legislative assembly and the parliament. It reviews the relief measures.
3. **District control room**- It monitors the rescue and relief efforts on continuing basis.
4. **Co-ordination**- The District magistrate also co-ordinates with the Central Government authorities and defence forces. He/she synchronizes the voluntary efforts of NGOs. The communication channels maintain through the police wireless network.

7.7 National Disaster Management Framework

Every citizen in this country is a part of National Disaster Management system that is all about section-protecting people and property from all types hazards. Think of the National Disaster Management frame work as a pyramid.

In the federal set-up of India, the responsibility to formulate the Government response to a natural calamity is essentially that of the concerned state of

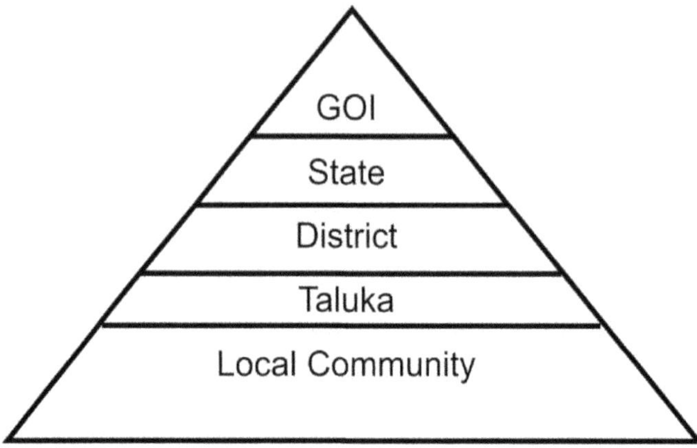

Government. The Government set-up/frame work for effective implementation of relief measures in the wake of national disaster is as follows.

7.7.1 Central Government

[Relief Commissioner *i.e.* Nodal Officer from Department of Agriculture and Co-operation (DAC)]

Committees at the national level

1. Cabinet Committee: Cabinet Secretary from Ministry of Agriculture.
2. National Crisis Management Committee (NCMC)

   Chairman: Cabinet Secretary

   Members: Secretaries from Prime Minister Office, Home Affairs defence, research and analysis wings.
3. Crisis Management Group (CMG):

   Chairman: Central Relief Commissioner

   Members: Senior Officers from various ministers and other concerned departments.

7.7.2 State

[Relief commissioner/Chief Secretary (Revenue department) is the head/in – charge of state administration for relief measures]

State Crisis Management Group (SCMG)

Chairman: Chief Secretary/Relief Commissioner

Members: Senior officers from the Revenue, Relief, Home, Civil supplies, power, Irrigation, Water Supply, Panchayat, Agriculture, Forests, public works and finance *etc.*

7.7.3 District

Focal point: Collector or Deputy Commissioner

7.7.4 Tehsil

Head: Tehsildar

7.7.5 Village (Local)

Contact person: village officer or patwari

7.8 Financial Arrangements

The dimensions of the response at the level of central government are determined in accordance with the existing policy of financing the relief expenditure and keeping in view the factors like:

1. The gravity of the natural calamity.
2. The scale of the relief operation necessary, and
3. The requirements of central assistance for augmenting the financial resources at the disposal of the state.

7.9 Community Based Organizations

The recent trends have revealed that the community based organization is the most powerful in the entire mechanism of disaster administration. Awareness and training of the community is particularly useful in areas which are prone to frequent disasters. The preventive action taken by community reduces the damage caused by the disaster. The disaster mitigation is most effective at the community level. The approach recognizes that when disasters occur, the communities of people themselves are the first responders. It is always the relatives, neighbours and other community members who come to help disaster victims and their families. The help from external agencies including the government, NGOs, political parties and private sector come later on.

It is laudable; the efforts in certain areas where communities have framed their own organizations which takes the right initiative in such situation. One such community based organizations is the Village Task Force formed in villages of Andhra Pradesh (South Indian State Prone to tropical cyclones) by the Church Auxiliary for Social Action (CASA). The village task force has been trained emergency evacuation and relief within the village. It is elected by the people themselves and during disaster it serves as the nodal body at village level which has mobilizes resources for the community and disseminates necessary information passed on by the outside agencies.

The community is an effective institution. The considerable efforts are being made to form and strengthen community based organizations at grass root levels.

7.10 The Armed Forces; Police and other Organizations

The armed forces of the country have played a vital role during disaster emergency, providing prompt relief of the victims even in the most inaccessible and remote area of the country. The organizational strength of the armed forces with their disciplined and systemized approach, and with their skill in technical and human resource management make them in dispensable for emergency situations.

Besides, when disaster occurs over large areas, it is usually beyond the capability of the administration to organize the relief activities, the armed forces are then called upon to organize the relief measures.

Related to the efforts of the armed forces, are the civil defence and the Home Guard Organizations. These organizations are voluntary in nature and character and come in handy emergency situations like natural disaster. A network of these is over the country. Their aim, while not actually taking part in actual combat operations like in army, Is firstly to save lives, to minimize damage to property and to maintain community of production. Thus, while disaster situations often led to chaotic conditions where rescue and relief work is severely affected. These organizations are able to co-ordinate and support efforts in a disciplined manner so that both the army and the district officials are able to carry out their respective activities efficiently.

Chapter 8

Physical and Socio-economic Impacts of Disasters

8.1 Introduction

· Disasters are no respecter of persons and the trail of destruction that they leave behind is a common occurrence. Their effect or impact, is usually felt across all sectors in society, at the community or individual level, which has led to push a for the more multi-sectoral approach to prepare and respond to disasters! The impact of a disaster may either be a direct or indirect one, its effect trickling into most homes and families in the community. The more obvious physical impact leads to the socioeconomic and emotional impact felt by the community. The intensity of the impact of any disaster is dependent on the preparedness level of the community or nation. Factors that increase the intensity of the effect of a disaster are poverty, environmental degradation, population growth, and lack of information and awareness about the hazards that exist in the area, and the potential risk they pose to the community at large.

8.2 Types of Impacts

Disasters impact heavily on the basic needs of people, and their livelihood, thus governments need to be prepared so that they can deal with the disaster promptly and effectively. Usually, immediately after a disaster has occurred, a team made up of government and non – government agencies is sent into the disaster site to carry out what is known as a Rapid Assessment exercise. The information collected from this quick initial assessment on the damage done from the hazard, is used by the leaders of the community or nation to determine whether any external assistance is needed. It is also used to determine the "what and how much relief" needs to be brought in immediately, and also what specific segments of society have been affected heavily by the hazard. The impact intensity felt by a community from a disaster is dependent upon the vulnerability of the community before the hazard struck (*e.g.* proximity to hazard, any education and awareness done etc) and thus their preparedness level. In any community the most obvious impact is the physical impact. The physical impacts in turn lead to the social impacts felt by the community. These are described in further detail below.

8.2.1 Physical Impacts

The physical impacts of a disaster are the deaths and injuries, and the damage to property and the built environment. The built environment can be classified as infrastructure and service sectors such as electricity, water *etc.* The amount of deaths can lead to a reduction in the population, and thus the workforce, which will in turn have an impact on the socio economic sector of the community. It should be noted here that the amount of physical damage caused by a hazard can affect the speed at which the response to the area can occur. If roads are cut off, this means alternative means need to be looked at to bring relief in to the disaster zone.

8.2.1.1 Infrastructure

Infrastructure includes the basic facilities, services and installation required for the functioning of a community or a society. Since these facilities, services and installations are spread throughout the community and country, they are normally impacted to some degree when disasters strike. Of the many components of a country's infrastructure, a select few are vital to both disaster response and to overall safety and security of the effected population. These components are referred to as "critical infrastructure," While all infrastructures damaged or destroyed in the disaster will eventually require rebuilding or repair, critical infrastructure problems must be addressed in the short term, while the disaster response is ongoing. The repair and reconstruction of critical infrastructure requires not only specialized expertise but also equipment and parts that may not be easily obtained during the emergency period. However, without the benefit of certain infrastructure components, performing other response functions may be impossible. Examples of critical infrastructure components include:

i. Transport System (Land, Sea and Air)

This system is important because at the time of disasters there needs to be an evacuation route available so as to get people out of the danger zone and or bring relief in. Transport is also important when a team needs to be sent in to the disaster zone to do a Rapid Assessment exercise. Transport mediums also need to be available, so that if one transport system is cutoff, another mode of transport can still be used.

ii. Gas and Oil Storage and Transportation

Connected to transportation above, there needs to be a store of the above to enable transportation of people out of the danger zone. Evacuation may take a couple of days to a week, and so extra fuel and oil is needed for the cars, boats, or helicopters etc that will be transporting people out.

iii. Communication

This is a critical because before a disaster and in the event of a disaster communication is needed. It is needed to get information out so that the outside "world" know what is needed and can respond appropriately.

vi. Electricity, Water Supply System, and Public Health

Damage to critical infrastructure which provide the above basic services needed by the community can affect the lives of people in the short term. In great need immediately after any disaster are water and sanitation, as well as the health of the disaster victims. Again, this is assessed in the Rapid Assessment exercise so that it is dealt with immediately.

v. Security

The management of past disasters was done on an *ad hoc* basis. As a result one of the many components overlooked was that of security, partly due to the fact that most of the resources were used for the immediate evacuation of people and saving lives! Today however security has become an important factor that has been mainstreamed into the action plans of many disaster planning offices. Security is the condition of being safe from harm, danger or loss. Security can be either emotional or physical security.

vi. Physical Security

Physical security is any and all necessary requirements that once implemented are designed to prevent, deter, inhibit or mitigate threats that face the safety and security of persons and/or property. Safety and Security in disasters differ by the fact that safety provides for the reduction of the risk of occurrence of injury, loss or death from accidental or natural causes. Security on the other hand provides for the reduction of the risk of occurrence of injury, loss or death from the deliberate or intentional actions of man and natural causes. Usually when disasters or an emergency situation arises, the following security issues arise:

a) Looting of Retail Outlets and Business Houses

Disasters or emergency situations provide an ideal opportunity for people to go on a looting spree. Looting arises especially when it has not been factored into the disaster emergency response plan or action plan. If no preparedness in this area has been done, when the hazard strikes to cause a disaster, most of the resources are being used to evacuate people. This leaves business houses and retail shops left unattended and vulnerable to looting by those looking for an opportunity for "free stuff". Looting may also take place after the immediate hazard has struck. This will usually happen when people have been waiting for some kind of response or assistance (recovery), and authorities have not been forthcoming with the needed aid. It is here that people say "Well we do not have any more jobs because everything was destroyed in the disaster, and so we do not have any money so how can we buy food?" As a result, looting takes place because of the fear that authorities will not take care of their needs, and so people find ways to take care of themselves and their families!

b) Security of Women and Children

Again if there is no preparedness, women and children are vulnerable to attacks of violence or rape by others, or even to the exposure to the primary hazard (fleeing to a danger zone) or secondary hazards; maybe because the lack of knowledge or panic. Violence or rape is more likely to arise if in the evacuation process, families have been separated from each other, and thus women and or children isolated from that security of their families. It is also more likely to arise in care centers where it is usually overcrowded.

c) Security of Aid Workers

There is now also the concern of the humanitarian workers who are flown into the disaster site to assist in the response of the disaster. Many humanitarian workers are foreigners to the site and so need to be aware of the hazards (human or natural) in the area and take necessary precaution. For women, there is also the security against violence or rape, especially in a war situation!

vii. Emotional Security

People have different emotional needs that when faced with disaster, will act differently depending on how serious the disaster is. When the physical needs of security, whether it is food security, physical security etc are not met, coupled with the fear of uncertainties, this can lead to stress and trauma, thus lack of emotional security.

8.2.2 Social Impacts

i. Welfare

Welfare falls into the socioeconomic and socio-political category. On the socioeconomic front this is represented by significant losses to Gross Domestic Product of the affected country or region. The local and national economy can experience low productivity, price slump, high unemployment and inflation. Small island states are more vulnerable compared to the larger developed nations when confronted with disasters of a large magnitude. There are overall financial impacts on the household and individuals that adversely impact on people's welfare for example dwellings, homes, property, and other assets can be damaged, sentimental value of assets can be lost forever which imply investment loss and reduction in the quality of life for the communities affected. For example people whose livelihoods depend on crops and livestock will face income loss that may impact on their welfare and overall wellbeing. For the business community (retailing, services, industries, wholesaling), their loss of income is represented by 'operational vulnerability' that is, the estimated time any business can operate without infrastructure support. For instance a business cannot operate without electric power (which is 0 hours), but it can operate for a maximum of 4 hours without phones. For time periods exceeding the above, the business ought to suspend operations indefinitely. Directly related to the immediate welfare needs of the victims/survivors are their food requirements. In this respect food assembling and distribution points have to be coordinated in such a way that is effective and efficient given the prevailing circumstances. Perhaps the welfare impacts of disasters can best be summed up by differentiating between the direct effects on property, the indirect effects brought about by the decline in factors of production, and secondary effects in the post disaster period such as economic decline manifested in balance of payments problems.

ii. Economic Impacts

Economic costs of disasters vary across space and time. Evidence suggests a strong correlation between a country's level of development and disaster risk. On average, 22.5 people die per disaster in highly developed countries, 145 people die per disaster in countries with medium human development and 1,052 people die per disaster in countries with low levels of development (UNEP). Sometimes the economic impacts can be difficult to calculate. The Western Indian Ocean islands experience more than ten cyclones a year between November and May: huge costs are incurred due to the destruction of income-generating activities including tourism revenues. To illustrate the financial loss to an economy, a study by SOPAC and the USP on the impacts of cyclone Ami on Fiji's agriculture sector estimated that 60-80 per cent of subsistence crops were damaged at a cost of FJ$921,000, direct damage to commercial crops was estimated at FJ$39.2m, the sugar industry lost 150,000 tones in lost production, and FJ$6m was lost in damaged infrastructure and equipment. In Bangladesh, floods during the Monsoon season destroyed crops and disrupted the non-farm economy of the country even after the flood waters receded. For instance, the average monthly working days fell in the period of the floods for farm workers. Day labourers for example were severely affected, their employment fell sharply from 19 days per month to 11 days per month, and as such, wage earnings also fell by almost 46 per cent. Another example is the great Hanshin earthquake in Japan in 1995 that caused $US 100 billion dollars damage which was equivalent to 2.1 per cent of Japan's GDP. Extensive damage to buildings, transportation facilities and utilities (gas, sewage, power) makes up 80 per cent of this cost. Recovery and reconstruction activities may start immediately after the event beginning with the most damaged sectors, whilst services sectors such as manufacturing may take up to twelve months or more for full recovery. It is understood by everyone that a community is referred to as the people who live in it. Out of the varieties of impacts, economic impacts are one of the major areas that need attention from the moment of any disaster. Just like food and shelter, education also needs to be included in the list of areas that contribute to the economic impacts. Due to the loss of household belongings and perhaps the parents, there will be a loss of the family income. In these situations most people tend to ignore the importance of education for all ages. This also happens as a result of evacuating from the home land to different regions. However, to get education, either the students need to be admitted to other schools, or should be provided with housing to come back. Whatever the conditions, education should not be interrupted as it is central to creating a new level of public awareness and preparedness. Attending school also keeps children away from parental issues back at home. Similarly, they will get opportunities to share their own situation with friends and elders at school which brings more liveliness for them. In addition to this, parents will find that their children are safe, and they too will have time to attend other activities. Hence, this is possible only if the community gets prepared to use the resources available in the community beforehand. Evacuation planning by using the available resources is a critical component of safety, including for people with disabilities. This is applicable for all buildings, including those that are new and fully accessible. Evacuation planning should include a need of assessment to determine who may need what in responding to an emergency and evacuating a facility. Also, this will inculcate the understanding of best use of the resources in the situation. In most places, during natural disasters and terrorism, in the development of preparedness plans school buildings will be the target of evacuation. In evacuation drills like fire in school buildings, the nearby parks and sports fields are allocated. In preparedness it is also very important for every member of the community to be familiar with the contact numbers during an emergency. This may include the numbers of Fire Station, Police Station *etc.* In this way the community will be aware of the use of the available resources rather than depending on the special resources provided during an emergency. Just like losing the household, death of children and adults also means the loss of future labour force, thus loss of future productivity. Hence, the regional economy and labour force is required to be maintained. This means for the recovery a labour force is needed to rebuild the infrastructure to replace places like houses, schools, utilities, *etc.* Whereas in severe disasters; there will be a higher unemployment rate than employment as the schools, factories offices are completely destroyed. This results in people having to attend to more than one job.

8.2.3 Emotional Impacts of Disasters

8.2.3.1 Introduction

In Chapter 1 you would have looked at the different types of disasters that occur. Whether it is a natural or a man-made disaster, the impact disasters have on the affected community, either collectively or individually, varies. All too often it is much easier to see the physical consequences of a disaster – injuries, death, and displacement. The immediate response to alleviate the pain and suffering is easily measured in terms of shelter, food, medicine, water and other things alike. Many victims of disasters have the ability to adapt to the sudden changes in their environment and daily routine more than others, and thus, are more resilient. The more prepared communities are before a disaster occurs, the more resilient they are. Today across the small island states, we are seeing the need to consider more the psychosocial effects that the disaster has on the affected communities; something that is less obvious. As research has shown disasters are expected to increase in the future. As a result, more people are expected to be emotionally and mentally affected, testing their resilience. There is a need for trauma counselling services in high risk areas, as well as the need to consider or put in place welfare services for those severely impacted by disasters! Programs to put in place trained counsellors at the community or village level have already been established in some areas.

8.2.3.2 Trauma

8.2.3.2.1 What is Trauma?

Anyone who goes through a disaster experiences some kind of trauma. For the less resilient communities or individuals, trauma can destroy them in that they cannot cope with the sudden event of a disaster and so they may suffer from a developing disease, lead to substance abuse, mental disorders and eventually destroy relationships, and the very fabric of society which are families themselves. Trauma is an exceptional experience in which powerful and dangerous events overwhelm a person's capacity to cope. When an adult or child is traumatized, they are experiencing reactions to the trauma that affect their ability to function. You must remember that when we talk about trauma, it is not only the survivors of a disaster that can be traumatized – there are also the relatives and friends of survivors, or emergency workers, and even seeing and hearing about disasters through the media (especially children!).

8.2.3.2.2 Factors of Vulnerability to Trauma

There are many factors which contribute towards one person experiencing trauma and not the other person. Many times it may not be just one particular factor that may be the sole cause or contributor, but more than one factor. In most cases people do get over the experience of surviving a disaster and move on with their lives. For others, it is easier said than done! Also, factors which contribute towards a child experiencing trauma may differ from that of an adult.

Generally, factors that may contribute towards trauma are:

a) Exposure or proximity to disaster site – Those who are closer to the disaster site are also going to feel more intense impacts and suffer more, than those who live further away from the disaster site.

b) Repeated images of terror on TV - this is especially true for children.

c) Relationships – Those whose relatives have been lost or injured in the disaster are at higher risk compared to those who have not lost anyone.

d) Age - older people are often less adaptable than younger community members.

e) History of previous traumatic events – If a person has gone through previous stressful events, whether it is violence, child abuse, etc, they are more likely to suffer trauma after a disaster has taken place as opposed to someone who has not.

f) Socio economic factors – generally, people who struggled to make a living before the disaster occurred, are more likely to be at more risk to trauma now that they may have lost everything, compared to someone who is more financially stable.

Can you think of any other factors?

8.2.3.2.3 Some Common Responses to Trauma

Common reactions to trauma may be any of the following:

- ✰ Shock
- ✰ Anxiety
- ✰ Fear
- ✰ Increased anxiety
- ✰ Guilt at surviving
- ✰ Sadness
- ✰ Confusion and
- ✰ Regret.

A trauma patient may also experience nightmares, lack of sleep, and flashbacks of the event. Physical reactions to trauma may be nausea, sweating, tiredness, loss of concentration, breathlessness, and aches and pains. In further response to all of the above symptoms, people may start to drink or smoke, substance abuse, throw themselves into work and or become anti social, avoid talking about what happened, as well as avoid any situation that may remind them of the disaster!

8.2.3.3 Counselling

8.2.3.3.1 What is Counselling?

The word counselling comes from the Middle English *counseil*, from Old French *conseil*, from Latin *cônsilium*; akin to *cônsulere*, to take counsel, consult. The definitions of counselling may vary in descriptions as different people see counselling in different perspectives. However, counselling can be defined as a relatively short-term, interpersonal, theory-based process of helping persons who are fundamentally, psychologically healthy resolve to developmental and situational issues. There are many different types of counselling which include disaster counselling, trauma counselling, cross cultural counselling, *etc.*

A good counsellor does the following:

- ✰ Listens effectively to what you are saying
- ✰ Works with you to define your goals with respect to your values and culture
- ✰ Facilitates your untangling of thoughts, feelings and worries about a situation
- ✰ Helps you gain your own insight into how you act, think and feel
- ✰ Teaches, shows and helps you express your emotions in your own way
- ✰ Teaches, shows and helps you work out your own solutions to problems
- ✰ Teaches, shows and helps you accept what cannot be changed
- ✰ Helps you become empowered to act in ways that are in your best interest
- ✰ Uses a variety of different techniques to help you explore what is important to you

8.2.3.3.2 Who Needs Counselling after a Disaster?

After a disaster most people go through a stressful time. Parents, children, heads of both the government and private sectors, and basically the whole community suffer some kind of trauma.

i) Children

In the early aftermath of disastrous events, many children encounter problems that are not easily resolved or their usual ways of handling problems are not working well for some reason. They may have found, for example, that talking to friends or relatives about their concerns is impossible or unsatisfying. Some of the concerns confronted by children include puzzling distressing feelings, low self confidence, getting along with others, self-defeating behaviors, academic problems, sexual identity concerns, and decision- making dilemmas. The Counselling Services can provide assistance for these concerns through counselling. Post-traumatic Stress Disorder (PTSD) is a psychological damage that can result from experiencing, witnessing, or participating in an overwhelmingly traumatic (frightening) event. Children often relive the trauma through repetitive play. In young children, upsetting dreams of the traumatic event may change into nightmares of monsters, of rescuing others, or of threats to self or others. PTSD rarely appears during the trauma itself. Though its symptoms can occur soon after the event, the disorder may appear months or even years later.

Some of these changes might appear in a child having PTSD.

- ☆ Refusal to return to school and "clinging" behaviour, including shadowing the mother or father around the house
- ☆ Persistent fears related to the catastrophe (such as fears about being permanently separated from parents)
- ☆ Sleep disturbances such as nightmares, screaming during sleep and bedwetting, persisting more than several days after the event
- ☆ Loss of concentration and irritability
- ☆ startled easily, jumpy
- ☆ Increase in behavioural problems, in school or at home in ways that are not typical for the child
- ☆ Physical complaints (stomach-aches, headaches, dizziness) for which a physical cause cannot be found
- ☆ Distance from family and friends, sadness, listlessness, less active, and preoccupation with the events of the disaster.

Professional advice or treatment for children affected by a disaster— especially those who have witnessed destruction, injury or death— can help prevent or minimize PTSD. Parents who identify any of the above changes in their children should take their children for a check-up.

ii. Adults

Usually adults do not hesitate to go to a professional counsellor because many adults and children find it helpful to talk to a counsellor specialized in post-traumatic reactions. Also it can be a cultural feature whereby it's not really a norm to talk about your problems in public or to confide to strangers. It is also important to consider that there might be severe cases among adults who might need assistance in getting to a counsellor to get diagnosed. Soon after a disaster it is important to have a group of counselors available for the victims to talk or attend to. However, if the number of counsellors is few compared to the population a selection of adult volunteers need to be trained, like teachers, council members, and parents. Being adults, teachers and trainers can play a wide role in helping their students by giving opportunities to come up with their experiences. Similarly, they can guide parents to help their children as well as themselves in the overcome of the trauma. Other than the people involved in schooling, counselling methods differ from culture to culture. How different cultural groups handle stress and deal with stressors, their abilities, needs and desires for certain types of assistance, their motivations, their sense of honor and pride, their religious orientations and beliefs, their political systems and leadership, and their ways of handling and dealing with grief and loss are just some of the variables which are affected by cultural differences. Therefore, it is important for the counsellors who are going to assist internationally in another culture to identify the important issues that will bring relief to the culture.

8.2.3.3.3 How Counselling can Help a Disaster Victim

Counselling gives the understanding of the actual situation in the perspective of the disaster victim. Hence different methods of counseling aid in adjusting to the environment. Counselling facilitates personal and interpersonal functioning with a focus on emotional, social, vocational, educational, health-related, developmental, and organizational concerns.

Some problems that can be helped by counselling:

- Coping with your relations to disaster.(fear, anger, coping with the changes in the environment)
- Exploring personal issues.(spirituality, sexuality relationships, your goals and ambitions)
- Family and relationship issues.(how to talk to the other person, intimacy with your partner)
- Dealing with practical issues(financial support, transport problems)

Populations served by counselling psychologists include persons of all ages and cultural backgrounds. Examples of those populations would include late adolescents or adults with career/educational concerns and children or adults facing severe personal difficulties. However, there is a lot of evidence that counselling can help you cope better with the many difficulties you face during and after a disaster.

8.2.3.3.4 How to find a Counsellor after a Disaster

When we say counselling, we are talking about the counselling people need just after an unusual occurrence such as the tsunami of 2004, severe hurricanes, cyclones and volcanic eruptions. Unlike the normal conditions (where anyone in very great distress can consult a doctor) after a disaster, anyone who is ready to listen to you is considered as a counsellor. The degree of medical knowledge of the counsellors or psychotherapists can be limited as it varies. Most counsellors work hand in hand along with the health facilitators.

Chapter 9

Terminologies Related to Disaster

Acceptable risk: The level of potential losses that a society or community considers acceptable given existing social, economic, political, cultural, technical and environmental conditions.

Adaptation: The adjustment in natural or human systems in response to actual or expected climatic stimuli or their effects, which moderates harm or exploits beneficial opportunities.

Biological hazard : Process or phenomenon of organic origin or conveyed by biological vectors, including exposure to pathogenic micro-organisms, toxins and bioactive substances that may cause loss of life, injury, illness or other health impacts, property damage, loss of livelihoods and services, social and economic disruption, or environmental damage.

Building code: A set of ordinances or regulations and associated standards intended to control aspects of the design, construction, materials, alteration and occupancy of structures that are necessary to ensure human safety and welfare, including resistance to collapse and damage.

Capacity Development: The process by which people, organizations and society systematically stimulate and develop their capacities over time to achieve social and economic goals, including through improvement of knowledge, skills, systems, and institutions.

Capacity: The combination of all the strengths, attributes and resources available within a community, society or organization that can be used to achieve agreed goals.

Climate change:

(a) The Inter-governmental Panel on Climate Change (IPCC) defines climate change as: "a change in the state of the climate that can be identified (*e.g.*, by using statistical tests) by changes in the mean and/or the variability of its properties, and that persists for an extended period, typically decades or longer. Climate change may be due to natural internal processes or external forcing or to persistent anthropogenic changes in the composition of the atmosphere or in land use".

(b) The United Nations Framework Convention on Climate Change (UNFCCC) defines climate change as "a change of climate which is attributed directly or indirectly to human activity that alters the composition of the global atmosphere and which is in addition to natural climate variability observed over comparable time periods".

Community disaster management organization: a national organization which ensures that planned activities for disaster management are implemented within a given timeframe

Community-based approach: a method of education and public awareness in disaster management in which community members are involved in the planning and implementation of the awareness programmes

Contingency planning: A management process that analyses specific potential events or emerging situations that might threaten society or the environment and establishes arrangements in advance to enable timely, effective and appropriate responses to such events and situations.

Corrective disaster risk management: Management activities that address and seek to correct or reduce disaster risks which are already present.

Corrosive: Something that will destroy or irreversibly damage a substance, including living tissue, by chemical action. The main hazards to people include damage to eyes, skin and tissue under the skin, but inhalation or ingestion of a corrosive can damage the respiratory and gastrointestinal tracts.

Critical facilities: The primary physical structures, technical facilities and systems which are socially, economically or operationally essential to the functioning of a society or community, both in routine circumstances and in the extreme circumstances of an emergency.

Development: A step or stage in growth or advancement in society, economics or in politics for a better lifestyle.

Disaster management cycle: A cycle with phases that reduce or prevent disasters

Disaster management: Is more than just response and relief (*i.e.*, it assumes a more proactive approach) Is a systematic process (*i.e.*, is based on the key management principles of *planning*, *organizing*, and *leading* which includes *coordinating* and *controlling*) Aims to reduce the negative impact or consequences of adverse events (*i.e.*, disasters cannot always be prevented, but the adverse effects can be minimized) Is a system with many components (*these components will be discussed in the other units*).

Disaster risk management: The systematic process of using administrative directives, organizations, and operational skills and capacities to implement strategies, policies and improved coping capacities in order to lessen the adverse impacts of hazards and the possibility of disaster.

Disaster risk reduction plan: A document prepared by an authority, sector, organization or enterprise that sets out goals and specific objectives for reducing disaster risks together with related actions to accomplish these objectives.

Disaster risk reduction: The concept and practice of reducing disaster risks through systematic efforts to analyze and manage the causal factors of disasters, including through reduced exposure to hazards, lessened vulnerability of people and property, wise management of land and the environment, and improved preparedness for adverse events.

Disaster risk: The potential disaster losses, in lives, health status, livelihoods, assets and services, which could occur to a particular community or a society over some specified future time period.

Disaster: A disaster is a situation in which the community is incapable of coping. It is a natural or human-caused event which causes intense negative impacts on people, goods, services and/or the environment, exceeding the affected community's capability to respond; therefore the community seeks the assistance of government and international agencies.

Early warning system: The set of capacities needed to generate and disseminate timely and meaningful warning information to enable individuals, communities and organizations threatened by a hazard to prepare and to act appropriately and in sufficient time to reduce the possibility of harm or loss.

Emergency: An emergency is a situation in which the community is capable of coping. It is a situation generated by the real or imminent occurrence of an event that requires immediate attention and that requires immediate attention of emergency resources.

OR "Is a situation generated by the real or imminent occurrence of an event that requires immediate attention" (key words) Paying immediate attention to an event or situation as described above is important as the event/situation can generate negative consequences and escalate into an emergency. The purpose of planning is to minimize those consequences.

Emergency management: The management of emergencies concerning all hazards, including all activities and risk management measures related to prevention and mitigation, preparedness, response and recovery.

OR The organization and management of resources and responsibilities for addressing all aspects of emergencies, in particular preparedness, response and initial recovery steps.

Emergency preparedness: Actions taken before the onset of a disaster so that a government can successfully discharge its emergency management responsibilities, such as establishing authorities and responsibilities for emergency actions and garnering the resources to support them.

Emergency services: The set of specialized agencies that have specific responsibilities and objectives in serving and protecting people and property in emergency situations.

Evacuation: Removal from hazardous place to another that is safe.

Flammable liquid: Any liquid that produces enough vapours to ignite if exposed to an ignition source.

Flammable solid: A substance when ignited, will burn so vigorously that it creates a hazard.

Hazard identification: The process of identifying what hazards have threatened a community, how often specified hazards have occurred in the past, and with what intensity (*i.e.*, damage-generating attributes measured by various scales) they have struck; the first level of hazard analysis sophistication.

Hazard Map: a map which shows areas that are vulnerable to particular hazards such earthquakes, cyclones, flooding, volcanic activity

Hazard: "Is the potential for a natural or human-caused event to occur with negative consequences". A hazard can become an emergency; when the emergency moves beyond the control of the population, it becomes a disaster.

Hazardous materials: 1. Any material that is dangerous to life, health, or property due to its chemical nature or properties. This group of chemicals is used in industry, agriculture, medicine, research, and consumer goods. Hazardous materials come in the form of explosives, flammable and combustible substances, poisons, and radioactive materials. These substances are most often released as a result of transportation accidents or because of accidents in chemical plants.

Hazardous waste: A solid waste, or combination of solid wastes, which because of its quantity, concentration, or physical, chemical, or infectious characteristics may: cause, or significantly contribute to, an increase in mortality or an increase in serious irreversible or incapacitating reversible serious illness, or pose a substantial presence or potential hazard to human health or the environment when improperly treated, stored, transported, or disposed of, or otherwise managed.

Humanitarian The act of promoting the welfare of humanity, especially through the elimination of pain and suffering.

Logistical readiness: A satisfactory state of readiness to mobilize resources in the most efficient and effective manner in order to minimize losses as a result of a disaster.

Logistics The branch of civil defence or agency that have to do with procuring, maintaining, and transporting materiel, personnel, and facilities

Memorandum of understanding: A legal document outlining the terms and details of an agreement between parties, including each party's requirements and responsibilities

Mitigation: The lessening or limitation of the adverse impacts of hazards and related disasters.

Poison: A substance that is toxic to life or health.

Public awareness: The extent of common knowledge about disaster risks, the factors that lead to disasters and the actions that can be taken individually and collectively to reduce exposure and vulnerability to hazards.

OR The process of transmitting information to the general population to increase their levels of consciousness about disaster risks so they can prepare appropriately to cope with a disaster

Reconstruction: A community or structure that has been reorganized, reformed, or restored after being impacted by a disaster or other hazard.

Recovery: The return of buildings and infrastructure to a normal or improved state after a setback or loss.

Rehabilitation: To restore buildings or parts of towns, to their former condition or better.

Relief: Private or public help in the form of money, food, clothing, shelter, or medicine, provided to people who are temporarily suffering from the effects of disaster and are at the time completely helpless.

Remittance: Sending of money to pay for resources or services to help people in need after a disaster.

Response mechanism: The means by which disaster relief is coordinated and mobilized from governmental and nongovernmental organizations to helpless victims of a disaster.

Response: Actions taken in reaction to a disaster or similar hazards.

Risk assessment: A methodology to determine the nature and extent of risk by analyzing potential hazards and evaluating existing conditions of vulnerability that together could potentially harm exposed people, property, services, livelihoods and the environment on which they depend.

Risk management: Consists of identifying threats (hazards likely to occur), determining their probability of occurrence, estimating what the impact of the threat might be to the communities at risk, determining measures that can reduce the risk, and taking action to reduce the threat.

OR The systematic approach and practice of managing uncertainty to minimize potential harm and loss.

Risk: "Is the probability that loss will occur as the result of an adverse event, given the hazard and the vulnerability" (key words) Risk (R) can be determined as a product of hazard (H) and vulnerability (V). *i.e.* R = H x V

OR The combination of the probability of an event and its negative consequences.

Security: Safety measures that provide a sense of protection against loss or harm from disaster or uncertain circumstances.

Volunteerism: The practice of using volunteer workers, especially in community service or disaster organizations and programmes.

Vulnerability: "Is the extent to which a community's structure, services or environment is likely to be damaged or disrupted by the impact of a hazard" (key words)

OR A condition wherein human settlements, buildings, agriculture, or human health are exposed to a disaster by virtue of their construction or proximity to hazardous terrain.

OR The characteristics and circumstances of a community, system or asset that make it susceptible to the damaging effects of a hazard.

Warning: Advice given to somebody or persons to be careful of impending danger.

References

Annual Disaster Statistical Review (ADSR). The numbers and trends.2014. Published by Centre for Research on the Epidemiology of Disasters.

"Bhopal Gas Tragedy: Fact Sheet". Hindustan Times. Dec 3, 2004.

Bankoff G, Frerks G and Hilhorst, D. 2004. Mapping vulnerability: Disasters, Development and People. Earthscan Publications, USA.

"Reducing Vulnerability to Natural Hazards: Lessons learnt from Hurricane Mitch", A Strategy Paper on Environmental Management, Stockholm, Sweden,1999 *http://www.iadb.org/regions/re2/consultative_group/groups/ecology_workshop_1.htm*. Accessed on 02/01/2016.

Broughton, Edward (2005). "The Bhopal disaster and its aftermath: a review". Environmental Health 4 (6): 6. *doi:10.1186/1476-069X-4-6.*

Dhavan, N.G. and Khan, A.S. 2012. Disaster Management and Preparedness.CBS Publications. New Delhi. India

http://earthquake.usgs.gov/earthquakes/eqarchives/year/byyear.php Accessed 02/01/2016 Largest and Deadliest Earthquakes by Year: 1990-2014

http://www.international.icomos.org/centre_documentation/bib/riskpreparedness.pdf Accessed on 02/01/2016

http://www.preventionweb.net/english/hyogo/gar/2015/en/garpdf/GAR2015_EN.pdf Accessed on 02/01/2015

http://www.preventionweb.net/english/hyogo/gar/2015/en/garpdf/GAR2015_EN. pdf. The 2015 Global Assessment Report on Disasters Risk Reduction (GAR15). Accessed on 02/01/2016

https://en.wikipedia.org/wiki/List_of_earthquakes_in_2015 on 02/11/2015

https://www.cfe-dmha.org/DMHA-Resources/Disaster-Management Reference -Handbooks Accessed on 02/01/2016 CFE-DM's Disaster Management Reference Handbooks

Human Security accessed on 15/11/15 from *http://humansecurity.gc.ca/*

Images are downloaded from:

http: / / www.sciencekids.co.nz / pictures / disasters.html

https: / / www.google.co.in / search?q=disaster+images+to+free+download and biw=1304 and bih=663 and site=webhp and tbm=isch and tbo=u and source=univ and sa=X and ved=0ahUKEwiN8OvslJfKAhUFjo4KHQz XCAUQsAQIOw#tbm=isch and q=nuclear+disasters and imgrc=YsutG1K 6yfdVnM per cent 3A

Kashiwazaki, T. (2005). Women in Disaster – looking back on the impact of 1995.

Johnson S, Sahu R, Jadon N, Duca C (2009). Contamination of soil and water inside and outside the Union Carbide India Limited, Bhopal. New Delhi: Centre for Science and Environment. In Down to Earth

Kobe earthquake on women. Retrieved on 20/11/15 from: *http://www.un.org/News/Press/docs/2004/pop911.doc*

Korstanje, M. (2011). "The Scientific Sensationalism: short commentaries along with scientific risk perception". *E Journalist*. Volume 10, Issue 2.

Kumar, A. 2014. Challenges to Internal Security of India. McGraw Hill Educations (India) Pvt. Ltd. New Delhi, India.

Menegazzi, C. 2004. "Cultural Heritage Disaster Preparedness and Response", Proceedings of the International Symposium held at Salar Jung Museum, Hyderabad, India, 23-27

Natural Hazard Risk Reduction in Project Formulation and Evaluation, *http://www.oas.org/dsd/publications/Unit/oea66e/ch02.htm.* Accessed on 02/01/2016

Phillips, B. D. 2005. "Disaster as a Discipline: The Status of Emergency Management Education in the US". *International Journal of Mass-Emergencies and Disasters.* Vol. 23 (1): 111–140.

Terminology on disaster risk reduction. 2009. Unite Nations International Strategy for Disaster Reduction (UNISDR), Geneva, Switzerland.

Risk Preparedness; Heritage at Risk, Bibliography, UNESCO-ICOMOS Documentation Centre, Paris.

Singh, J. 2007, Disaster Management: Challenges and opportunities, I. K. International Publishing House Pvt. Ltd. Bengaluru, Karnataka, India.

Singh, T. 2007. Disaster Management: Approaches and Strategies. Akansha Publishing. New Delhi.

Social Security accessed on 19/11/15 from *http://www.catgen.com/pekerti-foundation/EN/100000010.html*

Terminology of Disaster Risk Reduction, UNISDR;

http://www.unisdr.org/files/7817_UNISDRTerminologyEnglish.pdf. Accessed on 02/01/2016 UN International Strategy for Disaster Reduction, Geneva; *http://www.unisdr.org/archive/42814.*

Vaidyanathan. 2011. Introduction to Disaster Management. IKON Publications. Malaysia.

Whitehouse, Alfred, and Asep A.S. Mulyana. 2004. "Coal Fires in Indonesia." International Journal of Coal Geology 59: 91-97, 95.

Wood, K. (2005), Vulnerability of Women in Disaster Situations. On 20/11/07 from: *http://www.redcross.ca/article.asp?id=012396 and tid=001*

Appendices

Appendix I
Major Cyclones in World during 2005-2015

Name of Cyclone	*Year*	*Speed (km/hr)*
	North Atlantic Ocean	
Katrina	2005	280
Rita	2005	285
Wilma	2005	295
Dean	2007	280
Igor	2010	250
	Eastern Pacific Ocean	
Ioke	2006	260
Rick	2009	285
Celia	2010	260
Marie	2014	260
Odile	2014	220
Patricia	2015	325
	Western North Pacific Ocean	
Megi	2010	230
Sanba	2012	205
Haiyan	2013	230
Vongfong	2014	215
Soudelor	2015	215
	North Indian Ocean	
Gonu	2007	235
Sidr	2007	215
Giri	2010	195
Phailin	2013	215
Hudhud	2014	185
Nilofar	2014	205
Chapala	2015	215
	South West Indian Ocean	
Carina	2006	205
Hondo	2008	215
Edzani	2010	220
Bruce	2014	230
Hellen	2014	230
Eunice	2015	240

Contd...

Appendix I–*Contd...*

Name of Cyclone	*Year*	*Speed (km/hr)*
	Australian Region	
Floyd	2005	195
Glenda	2006	205
Monica	2006	250
George	2007	205
	South Pacific Ocean	
Meena	2005	215
Olaf	2005	230
Percy	2005	230
Ului2	2010	215
Pam	2015	250
	South Atlantic Ocean	
Anita	2010	85
Arani	2011	85
Bapo	2015	65
Cari	2015	65

Source: https://en.wikipedia.org/wiki/List_of_the_most_intense_tropical_cyclones.

Major Cyclones in India During 2005-2015

	Andhra Pradesh		
Yemyin	2007	Khai-Muk	2008
Laila	2010	Nilam	2012
Helen	2013	Lehar	2013
Hudhud	2014		
	Gujarat		
Yemyin	2007		
	Maharashtra		
Phyan	2009		
	Odisha		
Phailin	2013		
	Tamil Nadu		
Fanoos	2005	Nisha	2008
Jal	2010	Thane	2011
Nilam	2012	Madi	2013

Source: https://en.wikipedia.org/wiki/List_of_tropical_cyclones_that_affected_India.

Appendix II
Numbers of Major Earthquakes in the World during 2005-2015

Magnitude Ranging between	*2005*	*2006*	*2007*	*2008*	*2009*	*2010*	*2011*	*2012*	*2013*	*2014*	*2015*
8.0–9.9	1	2	4	0	1	1	1	2	2	1	1
7.0–7.9	10	9	14	12	16	21	19	15	17	11	19
6.0–6.9	140	142	178	168	144	151	204	129	125	144	123
5.0–5.9	1693	1712	2074	1768	1896	1963	2271	1412	1402	1577	1306
4.0–4.9	13918	12838	12080	12292	6805	10164	13303	10990	9795	14941	12608
Total	15762	14703	14350	14240	8862	12300	15798	12548	11341	16674	14057

Source: https://en.wikipedia.org/wiki/List_of_earthquakes_in_2015.

List of Major Earthquakes 2004-2015

Country	*Magnitude*	*Date*
Afghanistan	7.5	26 October 2015
Malaysia	6.0	5 June 2015
Sweden	4.7	15 September 2014
Thailand	6.3	5 May 2014
Japan	9.0	11 March 2011
Denmark	4.2–4.3	16 December 2008
Indonesia	9.2	26 December 2004

Source: https://en.wikipedia.org/wiki/Lists_of_earthquakes.

Major Earthquakes in India during 2005-2015

Location	*Year*	*Magnitude*	*Deaths*
North East India see 2016 North east India earthquake	January 3, 2016	6.7	6 dead, 100 injured in Manipur and Assam
Northern India, Pakistan, Afghanistan	October 26, 2015	7.7	260 in Pakistan and Afghanistan
Dibrugarh, Assam	June 28, 2015	5.6	0
Northern India, North East India	May 12, 2015	7.3	121+
Northern India, North East India	April 26, 2015	6.7[2]	Aftershock
Northern India	April 25, 2015	6.6[2]	Aftershock
Northern India, North East India	April 25, 2015	7.8[5]	8900+
Andaman and Nicobar Islands	March 21, 2014	6.7	0
Andaman and Nicobar Islands	April 25, 2012	6.2	0
New Delhi	March 5, 2012	5.2	1
Gangtok, Sikkim	Sept 18, 2011	6.9	118
Andaman Islands see 2009 Andaman Islands earthquake	August 10, 2009	7.7	26
Kashmirsee 2005 Kashmir earthquake	October 8, 2005		130,000

Source: https://en.wikipedia.org/wiki/List_of_earthquakes_in_India.

Appendix III
Major Fires in the World during 2005-2015

Country	*Year*	*Deaths*
Los Angeles and Ventura counties in California.	2006	–
Greek forest fires	2007	–
Camden Market Fire	2008	–
Victoria, Australia	2009	173
Kenyan	2009	111
Dhaka	2010	117
Manila, Philippines	2011	–
Queens, New York	2012	–
Yarnell Hill Fire	2013	19
Canada	2013	46
New Jersey, U.S.A	2013	–
Valparaíso wildfire in Chile	2014	13
Lahore supermarket fire in Pakistan	2014	13

Source: https://en.wikipedia.org/wiki/List_of_fires.

Appendix IV
Major Floods in the World during 2005-2015

Country	*Details*	*Year*	*Deaths*
China	Fujian, Anhui, Zhejiang flood	2005	1,624
India	Mumbai and the surrounding state Maharashtra, Karnataka, monsoon rain	2005	1,503
United States	2005 levee failures in Greater New Orleans	2005	69
Philippines	2006 Southern Leyte mudslide	2006	1,144
North Korea	2006 North Korea flooding	2006	844
Ethiopia	2006 Ethiopia flood, mainly Omo River Delta, Dire Dawa, Tena, Gode, flash flood, heavyrain	2006	705
Kenya, Ethiopia, Somalia	2006 East African Flood	2006	342
China	2007 China flood, mountain torrents, mud-rock flows	2007	1,348
North Korea	2007 North Korea flooding	2007	610
Mainly Sudan, Nigeria, Burkina Faso, Ghana, Kenya, and many African country	2007 African Nations flood	2007	353
Pakistan	2007 Balochistan flood by Cyclone Yemyin	2007	228
Indonesia	2007 Central and East Java torrential monsson rain, landslide, flood	2007	119
India	2008 Indian floods by monsoon rain	2008	2,400
South China	2008 South China floods	2008	200+
Taiwan	2009 August 8 flood, due to Typhoon Morakot, An entire village of Shiaolin was buried at the southern county of Kaohsiung	2009	677
Philippines	2009 Philippine Floods	2009	235–244
El Salvador	2009 El Salvador floods and mudslides	2009	199
Saudi Arabia	2009 Jeddah Torrential rain, floods	2009	123
Pakistan	Pakistan floods, monsoon flooding	2010	1600–2,000
China, North Korea	2010 China floods, landslides	2010	1,072+
Rio de Janeiro, Brazil	April 2010 Rio de Janeiro floods and mudslides	2010	246+
Colombia	2010 Colombia floods	2010	138
Jammu and Kashmir, Pakistan/India	2010 Leh floods	2010	125+
Rio de Janeiro, Brazil	January 2010 Rio de Janeiro floods and mudslides	2010	85+
Alagoas and Pernambuco, Brazil	2010 northeastern Brazil floods	2010	51+

Contd...

Appendix IV–*Contd...*

Country	Details	Year	Deaths
Salang tunnel, Afghanistan	2010 Salang avalanches	2010	172+
Asia	2011 Southeast Asian floods	2011	1,828
Philippines	Floods were caused by Tropical Storm Washi	2011	1,268
Brazil	January 2011 Rio de Janeiro floods and mudslides	2011	894
Africa	2010–2011 Southern Africa floods	2011	141+
North Korea	2012 North Korean floods	2012	223
Krymsk	2012 Russian floods	2012	172
Nigeria	Nigeria floods	2012	72+
India	2013 North India floods	2013	5,700
Greater La Plata, Argentina	2013 Argentina floods	2013	75+
India	Kashmir, J and K	2014	300-400
Serbia, Bosnia and Herzegovina, Croatia	2014 Southeast Europe floods	2014	80+
India	2015 Tamil Nadu floods Chennai, Cuddalore and Andra Pradesh named 2015 South Indian floods	2015	431

Source: https://en.wikipedia.org/wiki/List_of_deadliest_floods.

Annexure V
Most Air Polluted Cities in the World

Rank	*City and Country*	*PM 2.5 Pollution Level*
1	Delhi, India	153
2	Patna, India	149
3	Gwalior, India	144
4	Raipur, India	134
5	Karachi, Pakistan	117
6	Peshawar, Pakistan	111
7	Rawalpindi, Pakistan	107
8	Khormabad, Iran	102
9	Ahmedabad, India	100
10	Lucknow, India	96
11	Firozabad, India	96
12	Doha, Qatar	93
13	Kanpur, India	93
14	Amritsar, India	92
15	Ludhiana, India	91
16	Idgir, Turkey	90
17	Narayonganj, Bangladesh	89
18	Allahbad, India	88
19	Agra, India	88
20	Khanna, India	88

Source: http://www.politifact.com/truth-o-meter/statements/2015/oct/15/jim-webb/15-20-most-polluted-cities-world-are-india-china-s/.

Annexure VI
Top 10 Most Polluted Rivers in the World

Rank	*River*	*Country*
1	Citarum River	Indonesia
2	Ganges River	India
3	Mantanza-Riachuelo River	Argentina
4	Buriganga River	Bangladesh
5	Yamuna River	India
6	Jordan River	Israel
7	Yellow River	China
8	Marilao River	Philippines
9	Sarno River	Italy
10	Mississippi River	U.S.A

Source: http://listdose.com/top-10-polluted-rivers-world/.

Annexure VII
Disaster Management Authorities & Institutions in India

- National Center for Disaster Management, New Delhi
- National Information Center of Earthquake Engineering- IIT Kanpur, U.P.
- Disaster Management Institute, Bhopal, M.P.
- Disaster Mitigation Institute, Ahmedabad, Gujarat
- Environment Protection Training and Research Institute, Hyderabad
- Gujarat State Disaster Management Authority (GSDMA)
- Joint Assistance Centre, Gurgaon, Haryana
- Centre for Disaster Management (CDM), Pune, Maharashtra
- Sikkim Manipal University of Health, Medical and Technological Sciences, Tadong, Gangtok, Sikkim
- PRT Institute of Post Graduate Environmental Education & Research, New Delhi
- Central Board of Secondary Education, New Delhi
- Indira Gandhi National Open University, New Delhi
- National Civil Defence College, Nagpur, Maharashtra

Index

D

E

F

G

H

I

L

M

N

O

P

R

S

T

www.ingramcontent.com/pod-product-compliance
Ingram Content Group UK Ltd.
Pitfield, Milton Keynes, MK11 3LW, UK
UKHW021954270726
14060UKWH00002B/516

9 789386 071453